How to Become an Effective Course Director

Bruce W. Newton • Jay H. Menna
Patrick W. Tank

How to Become an Effective Course Director

 Springer

Bruce W. Newton
College of Medicine, Academic Affairs
Department of Neurobiology and Developmental Sciences
University of Arkansas for Medical Sciences
4301 West Markham St.
Little Rock, AR 72205, USA
newtonbrucew@uams.edu

Jay H. Menna
Emeritus Professor
College of Medicine
University of Arkansas for Medical Sciences
4301 West Markham St.
Little Rock, AR 72205, USA

Patrick W. Tank
Department of Neurobiology and Developmental Sciences
University of Arkansas for Medical Sciences
4301 West Markham St.
Little Rock, AR 72205, USA
tankpatrickw@uams.edu

ISBN: 978-0-387-84904-1   e-ISBN: 978-0-387-84905-8
DOI: 10.1007/978-0-387-84905-8

Library of Congress Control Number: 2008936628

Printed on acid-free paper

springer.com

Dedication

This book is dedicated to course directors who labor in the academic vineyards to ensure that their students receive the quality education that they deserve. We know how much work it is to direct a course and that is the reason that this text was written.

The authors would also like to thank the International Association of Medical Science Educators (IAMSE) for sponsoring the workshops on "How to Become an Effective Course Director", as these workshops form the backbone upon which this book was written. We would also like to thank the IAMSE for its dedication to the advancement of medical education.

Finally, we would like to thank our wives for their support of our professional endeavors.

BWN
JHM
PWT

Preface

Which Way Does Your Desk Face?

Early in my tenure as a medical neuroscience course director, I started receiving comments on student evaluations stating that I was "unapproachable". For the ten years prior to becoming the course director I taught full-time in the gross anatomy course and gave lectures in the medical neuroscience, histology and embryology courses. This amounted to over 130 student contact hours per year, during which time I had not received negative comments concerning "approachability". At the start of my third year as the medical neuroscience course director, I asked Dr. Patrick Tank, who was, and still is, the gross anatomy course director, why I was getting such comments. He looked up at me while I was standing in his office doorway and simply said, "Which way does your desk face?"

He then explained to me how he had arranged his office so that when he sat at his desk he faced the door to give students his immediate attention when they came to see him. My desk and chair faced the window, putting my back to the students. He stated that while my office arrangement avoided annoying reflections on my computer monitor, it sent an unintended message to the students that I did not consider them a priority. Once I moved my desk so that my chair faced the door the perception of me as being unapproachable was resolved.

This book is an outgrowth of a day-long faculty development workshop entitled, "How to Become an Effective Course Director", which was presented at several annual meetings of the International Association of Medical Science Educators (IAMSE). The topic for this workshop was conceived by Dr. Roger Koment, the founder of the IAMSE. The workshop was developed by Drs. Patrick Tank and Jay Menna who presented it twice before Drs. Bruce Newton and Robert Klein assumed direction in 2004 (Menna, Tank 2002, 2003; Newton, Klein 2004, 2005; Newton, Klein, Mylona 2006).

As the faculty development workshop evolved, the authors realized that the concepts we were teaching were applicable not only to faculty who direct medical science courses, but also to faculty members who direct courses in the allied health professions and courses at the college level. Thankfully, the editors at Springer also realized this and offered us the opportunity to put these concepts into book form.

The objective of this book is to make you the best course director and educator that you can be. To this end, the authors will share with you their collective experience of 38 years as course directors, 82 years as educators and 23 years as academic deans at the University of Arkansas for Medical Sciences (UAMS) College of Medicine. This text will address practical solutions to common problems encountered in becoming and remaining an effective course director. The hidden agenda is to return education to the centerpiece of institutions of higher learning.

The book has been divided into task-oriented sections to make your reading more focused and therefore more effective. We hope that in some small way this book will give you direction so that you will not have to "learn the hard way" as the authors did. We welcome your comments as to the effectiveness of this book as well as the sharing of your experiences as a course director and educator.

The workshop and book would not exist if it were not for the support of the IAMSE. The IAMSE is currently the only international organization whose sole purpose is to enhance the educational skills of basic science faculty in all medical professions. The IAMSE is truly an international organization. As of 2008, IAMSE has over 700 members from 40 countries who meet annually to share ideas with the objective of helping each other improve our educational and course leadership skills. Faithful to our mission statement, "To advance medical education through faculty development and to ensure that the teaching and learning of medicine continues to be firmly grounded in science", IAMSE takes pride in developing "teachers" into "educators". Our motto is "concept to classroom", and indicates that we are dedicated to the practical application of educational techniques, technology, and theory by educators in the classroom. For more information about IAMSE, or for information on how to join, please visit www.iamse.org. As a current member of the Executive Board, I am pleased to invite each reader to join IAMSE – we would be more than pleased to have you contribute your skills and knowledge to our organization.

In closing, I would be remiss if I did not mention that Patrick Tank remains a friend and a mentor, as does Jay Menna and that much of the success I enjoy as an educator and an Associate Dean I owe to them. The authors don't have all of the answers about how to direct a course but we have many of them, and we are willing to share with you our experience and our insights. Now let's get started.

Little Rock, Arkansas   Bruce W. Newton
June, 2008

About the Authors

Dr. Bruce Newton received a B.S. in biology from Slippery Rock University, Slippery Rock, PA in 1978. He received his Ph.D. in Anatomy from the University of Kentucky, Lexington, KY in 1984. He did a four year post-doctoral fellowship in Neuroscience at the University of Rochester, Rochester, NY before taking a position in 1988 in the Anatomy Department at the University of Arkansas for Medical Sciences, Little Rock, Arkansas — he has been there ever since. Dr. Newton has taught in the Gross Anatomy Course since 1988, and during that time was the Medical Neuroscience Course Director from 1996 – 2003. In January, 2007 he accepted the position of Associate Dean for Undergraduate Medical Education. In 2001 he joined the International Association of Medical Science Educators (IAMSE). Since 2003, he has been elected for three consecutive terms as a member of the IAMSE Board of Directors and has been the treasurer since 2004. Dr. Newton has won several teaching awards. His educational research interests include changes in vicarious empathy and critical thinking skills during undergraduate medical education. His basic science research looks at the development of the autonomic nervous system and sex differences in pain.

Dr. Jay Menna received a B.S. Degree, *Magna Cum Laude* from Jacksonville University, Jacksonville, FL, in 1970; a M.A. Degree from the State University of New York (SUNY) at Buffalo, in 1972; and his Ph.D. in Microbiology and Immunology from SUNY at Buffalo in 1975. From 1975 to 1976 he was a post-doctoral fellow at Vanderbilt University, Nashville, TN. He then accepted a position as Assistant Professor of Microbiology and Immunology, Department of Microbiology and Immunology, College of Medicine, University of Arkansas for Medical Sciences (UAMS), Little Rock, Arkansas, and has been at UAMS since then. During that time Dr. Menna was the director of the Medical Microbiology course for five years. From 1989 to 2001 he served as the Assistant Dean for Medical Education and in 2001 was promoted to Associate Dean for Undergraduate Medical Education until 2007 when he retired. Dr. Menna is a founding member of the International Association of Medical Sciences Educators (IAMSE), and served as an Editor of the Basic Science Educator, IAMSE; and as the Chair of the Planning Committee of this organization as well. In 2007, he was elected as the first Emeritus Member of the IAMSE. Dr. Menna has won numerous teaching awards and has published various articles and abstracts in the area of medical education. He still teaches in the College of Medicine, UAMS and was promoted to Emeritus Professor of Microbiology and Immunology, in November of 2007.

Dr. Patrick Tank received a BS Degree, *Cum Laude* in biology from Western Michigan University in 1972, and a MS in Anatomy from The University of Michigan in 1973 before completing his Ph.D. in Anatomy at The University of Michigan in 1976. He spent two post-doctoral years at the Developmental Biology Center, University of California, Irvine before joining the faculty of the Department of Anatomy at the University of Arkansas for Medical Sciences in 1978. He has risen through the ranks and is currently Professor of Neurobiology and Developmental Sciences (formerly known as the Department of Anatomy). He is the Course Director of Gross Anatomy and the Director of Medical Education for the Department of Neurobiology and Developmental Sciences. Dr. Tank has received numerous teaching awards, including the Golden Apple Award (winner 5 times, runner-up 12 times), the Scarlet Sash Award (20 times) and the Gold Sash Award (2 times). In 1996 he was awarded the Distinguished Faculty Award by the UAMS Caduceus Club, and in 1998 he became the first holder of the Charles Hartzell Lutterloh and Charles M. Lutterloh Medical Educational Excellence Professorship.

Table of Contents

1

What *Is* A Course Director?

This question is best addressed by asking another question - what does a course director do? The following is a list of responsibilities of a course director presented in a temporal sequence beginning with the development of course concepts and the setting of course goals. A course director:

- Develops course concepts and sets course goals
- Develops a course schedule
- Prepares the course syllabus and other supporting materials
- Selects text materials in support of the course concepts and goals
- Recruits course faculty
- Guides course faculty to achieve the course goals
- Supervises the construction of student assessment materials (examinations)
- Administers examinations
- Collects and archives the examination outcomes (keeps the grade book)
- Councils and assists students
- Reports student outcomes to the administration (submits the grades)
- Prepares and administers course evaluation materials
- Reviews course outcomes and makes modifications for the next year
- Repeats all of the above

As can be seen, the tasks of a course director are varied and at times require a good understanding of group dynamics. As we work our way through this book, we will elaborate on the task areas bulleted above. It should be obvious from the outset that it takes time and patience to become a dynamic and effective course director.

B.W. Newton et al., *How to Become an Effective Course Director*,
DOI 10.1007/978-0-387-84905-8_1, © Springer Science+Business Media, LLC 2009

2

Whose Idea Was This?

Clearly, every course needs a director and the assignment of that task is the ultimate responsibility of a department chairperson. How you become a course director occurs via two general routes:

2.1 It Was My Idea!

Yes, there are faculty members who want to become a course director. In the view of the authors this is the best scenario because the desire to be a course director frequently goes hand-in-hand with the desire to be an effective course director. One can rightly ask – why would anyone want to become a course director? There may be financial rewards; but many times the desire to become a course director reflects the personal satisfaction that one derives from teaching and interacting with students. This desire will be readily noticed and appreciated by students and by your teaching faculty.

2.2 It Was *Not* My Idea!

During the course of our workshops the authors have come to realize that many times a faculty member is told that they WILL be a course director. This conscription route to becoming a course director is usually based on need. Obviously for the non-volunteer, becoming a new course director is stressful and diaphoretic at the very least! The faculty member's response to this new stressor can be either positive or negative, with the latter response often being reflected in the interaction of the course director with his/her students, their teaching faculty and with academic administration. For the sake of the students, the course director's attitude must become: "I can do this and I can do it well ".

If you were conscripted, the authors hope that you have developed a positive approach to your responsibilities as a beginning course director. A positive attitude will make your course directorship rewarding as you witness the learning of your students and the satisfaction of your teaching faculty, your chairperson and the administration.

B.W. Newton et al., *How to Become an Effective Course Director*,
DOI 10.1007/978-0-387-84905-8_2, © Springer Science+Business Media, LLC 2009

3

Anatomy of a Course Director

At the beginning of our workshops, the first step is always to open the meeting with a discussion of the important characteristics of a course director. In our workshop experiences we have noticed that there may be differences in how different groups view the task of course director. Beginning course directors (those who are preparing for their first year as a course director) are typically a little naive, whereas veteran course directors are a little cynical. As our starting point in this book, it is important that you know what *you* think are the most important characteristics of a course director. So, let's begin with a self-test.

3.1 What Do *You* Think?

We have listed below six attributes that we believe are important characteristics of a course director. Please take a moment to rank-order this list. We will then show you what other beginning and veteran course directors think are the most important characteristics and then reveal what students think are the most important attributes of a course director.

A good course director:

_____Is fair

_____Has good communication skills

_____Is sensitive to student needs

_____Is flexible

_____Is enthusiastic

_____Has good organization skills

Scale: 1 = Most important attribute

6 = Least important attribute

B.W. Newton et al., *How to Become an Effective Course Director*,
DOI 10.1007/978-0-387-84905-8_3, © Springer Science+Business Media, LLC 2009

3.2 What Beginning Course Directors Think — Taking Their Pulse!

During our faculty development workshops, the authors have surveyed 85 beginning course directors as a means of determining what they think are the most important characteristics of an effective course director. These course director candidates were mainly from allopathic medical schools, with a smaller contingent from osteopathic medical schools and a chiropractic school. Regardless of the year that the survey was performed, or the institution of origin of the respondent, the responses that we received from beginning course directors break down in a similar fashion. The rank-order given to these six attributes by beginning course directors is shown in Table 1.

Table 1
What Beginning Course Directors Think Are the
Most Important Attributes of a Course Director

A good course director:

2	Is fair
4	Has good communication skills
3	Is sensitive to student needs
6	Is flexible
5	Is enthusiastic
1	Has good organization skills

Scale: 1 = Most important attribute
6 = Least important attribute

Beginning course directors also responded to an open-ended question about what they think are the most important attributes not listed in the survey. The most common responses in order of frequency were:

- Should have a comprehensive knowledge of course content
- Should have integrity and dignity
- Should have patience with disinterested colleagues
- Should keep the goals of the course in mind
- Should know where the course fits into the entire curriculum

It is of interest to note that many beginning course directors thought that a comprehensive knowledge of course content should have been included in the survey list of six attributes, since they felt this was something that students would consider a high priority. This was not the case, as will be discussed later.

3.3 What Veteran Course Directors Think — Taking *Their* Pulse!

During our workshops, 25 faculty members who were veteran course directors in allopathic, osteopathic or chiropractic schools were asked to rank-order the characteristics that a course director should possess. The responses of veteran course directors did not vary by year of survey or the type of institution where they were employed. The rank-order given to the six attributes by veteran course directors is shown in Table 2.

Table 2
What Veteran Course Directors Think Are the
Most Important Attributes of a Course Director

A good course director:

3	Is fair
1	Has good communication skills
4	Is sensitive to student needs
6	Is flexible
5	Is enthusiastic
2	Has good organization skills

Scale: 1 = Most important attribute
6 = Least important attribute

In addition to the survey results, veteran course directors also responded to an open-ended question about what they think are the most important attributes not listed in the survey. The most common responses in order of frequency were:

- Should have a thorough knowledge of course subject matter
- Should be willing to consider new teaching innovations
- Should have skill at constructing examinations that appropriately reflect course content
- Should be able to work effectively with others

Much like beginning course directors, veteran course directors thought that a thorough knowledge of course material was a highly desirable attribute. Veteran course directors would have rated that attribute very highly if it was included in the survey.

3.4 What Students Think — The Consumers Speak Out!

Over 200 students from allopathic, osteopathic and chiropractic schools were surveyed. These students were in their basic science years of education. Approximately 66% of the respondents were from five allopathic schools, 22% were from two osteopathic schools and 12% from a single chiropractic school. The students were asked to rank-order the most important characteristics that a course director should possess in descending order of importance. The responses did not vary by year of survey or the type of institution that the students attended. The rank-order given to these six attributes by students is shown in Table 3.

Table 3
What Students Think Are the
Most Important Attributes of a Course Director

A good course director:
1   Is fair
2   Has good communication skills
4   Is sensitive to student needs
6   Is flexible
5   Is enthusiastic
3   Has good organization skills

Scale: 1 = Most important attribute
6 = Least important attribute

Noteworthy about these results is that fairness is rated number 1 by students but is either ranked number 2 or 3 by faculty respondents. When asked about the fairness issue, faculty taking the workshop stated that fairness is considered a requirement for the job and is not ranked at the top of the list for that reason. We will have more to say about fairness later in this book

A second outcome that surprised the authors, is that all three groups ranked flexibility at the bottom of the list. Faculty want to set the goals and not yield. Students want the goals to be identified, to remain stationary and be attainable.

In addition to the survey results, students also responded to an open-ended question about what they think are important attributes or activities not listed in the survey. Their responses are listed, in no particular order in Table 4.

Table 4

Other Attributes of a Course Director Desired by Students

- The course director should attend all lectures so examinations reflect course content
- The course should have clear lecture and course objectives
- The course director should have enough time to do his/her job effectively
- The course director should make material relevant and cover core content for licensing/board related examinations
- The course director should have a sense of humor
- The course director should use a variety of teaching methodologies for different learning styles
- The course director should accept constructive criticism gracefully
- The course director should expect student excellence
- The course director should use reasonable reading assignments
- The course director should provide numerous practice problems/questions

Many of these desirable attributes are discussed elsewhere in this book, but some of these attributes warrant comment here since they are not mentioned in other chapters.

3.4.1 The Course Director Should Make Material Relevant

The authors agree. When possible, place lecture topics into a framework that makes the material applicable to the students' future career goals. Students want to know why they have to learn the material and how it will be useful to them in their careers. Students want core content to be taught and effectively explained to them. Students will then be able to flesh-out and solidify core content when they study the material after the lecture. Currently, many medical students have laptop computers and many have PDA's that are linked to sites that keep them up-to-date on the details of a particular topic. That's why didactic lectures that simply list details are less effective, whereas explaining the basic principles underlying the core content has greater intrinsic and formative value to students. Pursuant to this, the Liaison Committee on Medical Education (LCME), the organization that accredits U.S. and Canadian medical schools, also recognizes that memorization of facts does not necessarily produce a competent physician and now requires that students must be taught critical-thinking and lifelong learning skills as part of the core curriculum (LCME 2008).

3.4.2 The Course Director Should Have a Thorough Knowledge of Course Subject Matter

How knowledgeable does a course director have to be concerning course content? In response to an open-ended question, both beginning course directors and veteran course directors considered a comprehensive knowledge of course content a highly valued attribute. This is a misconception, since students didn't mention this attribute when asked the same open-ended question. As illustrated in Table 4, the students want to understand core concepts and would like those concepts taught in a relevant manner.

3.4.3 The Course Director Should Have a Sense of Humor

Students appreciate it when course directors have a sense of humor. They want you to be able to laugh at yourself when you make an error. An occasional unintentional double entendre usually evokes laughter from the class and you should be confident enough in your skills to be able to laugh with them. However you must exercise caution when intentionally using humor, since it is almost guaranteed that someone will be offended, no matter how innocuous you think your humor may be. The safest route to take is simple: if there is any doubt in your mind about whether or not you should say something, *do not say it*.

3.4.4 The Course Director Should Use a Variety of Teaching Methodologies

Teaching styles are one of the most investigated areas of educational research (Barrows, 1994; Stinson, Miller 1996; Hark, Morrison 2000; Hudson, Buckley 2004; Michaelsen, Knight, Fink 2004; Shanley 2007). Each student is an individual and has his/her own preferences about how he/she would like to be taught. Some students are auditory learners and find attending lectures an efficient learning modality. Others are visual learners and learn more efficiently by observing varying types of images. At the other end of the spectrum are students who simply want to be left on their own in terms of their learning. Most students are "hybrids" of these methodologies.

Experience has taught the authors that most students want information reduced to tables or easily handled small information bites. Many students learn efficiently by using well-organized computer-based materials and learning modules (Draves, 2002). The best advice we can offer is to explore the effective teaching styles that have been used in your discipline, and use as many different styles as possible as a means of addressing the multiple learning styles of your students. If your institution has an Office of Educational Development (OED), they will probably have a battery of tests that a student can take to define what type of learner he/she is.

It is generally not possible to accommodate all learning styles. Therefore, didactic lectures must be as effective as possible. For those students who learn better by reading, make sure that each topic has a concise reading assignment. If possible, develop computer-based, self-directed learning modules for the most important core content as a means of augmenting lecture material (Draves, 2002). Many students gauge their progress by self-assessment and your students should be encouraged to do so. Having a bank of up-to-date study questions will enhance this process and your students will appreciate your efforts in this regard. Ultimately, regardless of how the information is presented, the onus of mastering the material rests on the student. Therefore, let's try to make it as organized and easy for the student as possible.

3.4.5 The Course Director Should Expect Student Excellence

A number of students indicated that they wanted course directors to promote student excellence. It may seem surprising that students would want a challenging course. However, it makes sense that students who can excel do not want less qualified students obtaining the same grade for mastering a limited amount of material. The authors agree with the students. It is the responsibility of the course director to assemble a thorough course and develop examinations that discriminate the better students from the poorer students. In doing so, you are fulfilling your responsibility to your administration and to the profession for which you are training the future work force.

4

Living With Our Differences

Can a course director reconcile students' expectations with their own expectations? Let's get real here – we can strive for consonance with student expectations of a course director but, to be effective, we cannot comply completely with their wishes – there will always be differences. It is the hope of the authors that these differences are minor.

There are characteristics of a course director that the authors think are so important that they should not be subject to modification based on student pressure. We have taken the liberty of discussing them below.

4.1 That's Not Fair!

What is fair? According to Webster: "Fair (adj.) — marked by impartiality and honesty: free from self-interest, prejudice or favoritism; conforming with the established rules."

Fairness is the highest ranked trait of a course director from a student point of view (see Table 3). What is fairness in the eyes of medical students? Fairness is the equal application of the rules and regulations of a course, with some jurist prudence added for good measure. Fairness is assigning a reasonable amount of homework or reasonable reading assignment as compared to concurrently running courses. Fairness is not delivering too much new material immediately before an examination. Fairness is adhering to the allotted lecture time to preserve the break time between lectures. Fairness is having review sessions that are open to all students. Fairness is being consistent when hand grading fill-in-the-blank or essay questions. You get the idea.

Beginning and veteran course directors do not give as much thought to fairness as do students (they ranked this attribute second and third, respectively, see Tables 1 and 2). For course directors, fairness is treating all students equitably. There must be no equivocation in applying the same rules, regulations and protocols in all situations to all students. Any deviation from established guidelines, unless it is applied to all, will be construed by students as favoritism towards the student(s) for which the exception was made. Therefore, making certain that everyone receives the same treatment will prevent student dissatisfaction. If you remember to apply the rules and not interpret them, you will save yourself considerable grief. A course director should not put himself in a position wherein a student can go to

B.W. Newton et al., *How to Become an Effective Course Director*,
DOI 10.1007/978-0-387-84905-8_4, © Springer Science+Business Media, LLC 2009

the administration and prove that some students were treated differently than others.

The authors have occasionally heard a student say that they are being treated unfairly. Our response has been to ask the student if we (the faculty or administration) are doing anything different to them than what we are doing to their classmates. Their answer is always "No", and this often defuses the situation since that is the definition of equitable treatment.

4.2 Good Organizational Skills

Beginning and veteran course directors ranked good organizational skills as the first or second most important characteristics of a competent course director, respectively (Tables 1 and 2). Students ranked this attribute third on their list (Table 3).

Although being organized and paying attention to detail was not ranked at the top of the student list, the authors are pretty certain that poor organization will guarantee a poor course evaluation and harsh comments from the students. Good course organization begins before the course starts and continues throughout the course.

Good organizational skills demand daily, and sometimes hourly, attention to detail. Procrastination is not an option. Informing lecturers of the location of their presentation(s), immediately informing students of schedule changes, and preparing examinations on time are activities that require good organizational skills. Often faculty will need answers regarding course or university policies in order to respond in a timely fashion to questions from students, other faculty members, and administrators. Having the resources close at hand and being able to access that information is an essential organizational skill for the course director.

Some situations undermine the appearance of course organization. These include making numerous changes in the course schedule, having to verbally correct typographical errors on an examination while the examination is being administered, failure to post grades in a timely fashion, failure to counsel faculty on lecture organization, and failure to respond in a timely fashion to emails and telephone messages. Disorganization is anathema to students and will dramatically reduce student satisfaction with your course and can negatively impact student learning.

4.2.1 Organize Before Your Course Starts

A "trigger calendar" is a document that details a timeline for certain aspects of preparing and delivering the course. It is invaluable. For example, you will probably be asked to submit a lecture schedule, syllabus, and lecture/lab room requests on deadlines that fall far in advance of when your course starts. Knowing when your college needs this information will keep you on track and enable you to avoid scheduling conflicts with other courses and colleges within your institution. Develop a trigger calendar, or even better, inherit one from the previous course director.

If your course is team-taught (i.e., taught by multiple faculty members), it is probable that you will need to ask your faculty to prepare various aspects of the course, e.g., their lecture outlines or lab manual materials. Place your deadlines for receipt of these materials on your trigger calendar and periodically check with faculty members as your deadline approaches.

When preparing a syllabus, it is wise to give yourself about four weeks lead time before the finished product is to be printed. Accordingly, you will need to add this lead-time to the deadlines for your faculty members. This will give you time to check the syllabus for completeness before it is submitted for printing, and will allow time for delays in receiving the syllabus materials from your teaching faculty.

4.2.2 Organize the Course Calendar

Prepare a well-organized course calendar that contains the appropriate information about date, time, lecture topic, lecturer, and location of lecture. Be sure to include the examination dates, times and rooms. Once this schedule is set and distributed, be reluctant to modify it.

Students get very stressed if the course calendar that is printed in the course syllabus does not match what is actually occurring in the course. Although a course calendar is usually printed at the beginning of the year, the students must understand that changes can occur and that the electronic version of the calendar (if available) will have the most up-to-date information. Ultimately it is the re-sponsibility of the course director to make sure that all students are informed of any schedule changes.

If possible, build a course website and have a link to the policies of the institu-tion that include the student, faculty and course director handbooks as well as a course calendar that is updated frequently. Make sure that a process exists that will ensure that all changes in lecture topics and times are communicated to you as a course director, so that you are kept "in the loop".

If your institution is located in a geographic area where inclement weather is common, it is wise to include several "snow days" into your course calendar. Plan no new material during these times, but schedule the rooms/laboratories in case a class is cancelled and needs to be rescheduled at a later date. Having a few snow days inserted into your course schedule will allow you to adjust the lecture times and, hopefully, avoid having to cancel lectures or postpone an examination. If the weather is good, and the snow days are not needed for lectures or exams, these time slots can be used for review sessions for upcoming examinations.

4.2.3 Be Vigilant to Stay Organized

Good course organization requires that a course director remain vigilant while his/her course is in progress. It is essential that you inform your lecturers of the day, time, place, and topic of their lecture(s). Send a reminder one week and one day before the lecture session is to be given. Make sure that you have the correct email addresses, phone numbers or pager numbers of the entire teaching faculty in your course. If the lecturer is a clinician it may be more efficient to communicate with his/her administrative assistant, for the assistant is most likely to ensure that the information is included on the clinician's calendar.

Check on course-related information at least twice per work-day. Make certain that you respond to emails and telephone messages in a timely fashion. Responding in a timely fashion will make for satisfied faculty and students. When an examination is to be given, it will be necessary for you to check your email messages and phone messages the day before and the day of the examination because students may attempt to contact the course director to say they will be absent and will want to know if their reason for missing the exam is valid.

4.3 Sensitivity to Student Needs

Students rank being sensitive to student needs as fourth on their rank order list (see Table 3). The beginning and veteran course directors were in close agreement, ranking this trait as third and fourth, respectively (see Tables 1 and 2). Even though this trait ranks down the list, the experience of the authors shows that students want to have their opinions and comments taken into consideration. Therefore, it is imperative that course directors be approachable by students. In fact, the authors are of the opinion that this attribute is *sine qua non* in becoming and remaining an effective course director. Please let this be your credo when directing your course. Indeed, the preface to this book, "Which Way Does Your desk Face?" is about being approachable.

Student opinions and comments fall into the two broad categories – constructive and non-constructive. Constructive comments are good things and as a course director you should keep an open mind and respond to them. Let the students know that you have read their comments and have taken them into consideration. Inform the students which comments are useful and explain why others are not educationally sound. Non-constructive comments are not necessarily bad things. Students need to vent occasionally. Sometimes, you will receive caustic comments about your course, yourself, or your teaching faculty. As a course director you are the "lightning rod" for anything that the students consider worthy of criticism. Don't take it personally. The philosophy of the authors is quite simple: "When a course goes well everyone gets credit, but if something goes wrong the course director assumes the blame."

There are some things that a course director should not do.

- Down playing or dismissing the comments or problems of students who come to your office as unimportant or unworthy of your attention
- Being arrogant
- Being pompous
- Reprimanding the whole class for the actions of a few students

In regard to the last point, we are not saying that you should never reprimand a class. For example if the class does not perform well on an examination they should be informed but acknowledgement of their poor performance should be immediately followed with comments indicating that they have the capability to do better and that you and your faculty will assist them in their efforts. Another way to get a warning to the class is to inform the class officers of your concerns and have them relay the information to the class. Students are usually more willing to listen to their peers than to authority figures when the topic involves behavioral issues.

5

Where Do I Start?

There are many things you must do to ensure the smooth operation of your course. Many of these must be done before the course starts. Begin organizing your course at least two months in advance of the time it starts.

5.1 The Curriculum

One of the most important things that a course director should know is the position of his/her course within the overall curriculum of their institution. In a medical school setting this is extremely important given that there are few elective courses and elective clinical rotations until the students' senior year. In short, the curriculum is very structured, and in most cases highly coordinated both vertically and horizontally across each year and each successive year, respectively. What is taught in a given course is based on what has been taught and what is to be taught. Beginning course directors would be well served to meet with their academic dean(s) and other course directors to learn the temporal and topical relationship of their course within the context of the overall curriculum. The importance of this process becomes even more acute when a beginning course director assumes the directorship of a course that is presented within an organ-system modality or other form of tightly integrated curriculum.

Organize and schedule course content so information flows naturally from one topic to the next. This is especially true if your curriculum is integrated with other concurrent courses. Inserting lecture topics at the convenience of the lecturers often disrupts the normal progression of the material and disjoints student learning. To help avoid this, it is a good practice to know when major meetings are occurring in the various research disciplines of your teaching faculty. Meeting schedules are posted several years in advance, and referring to them will enable you to appoint a new lecturer to a topic when the faculty member who traditionally gives the lecture is going to be off campus. Doing this very early in the scheduling process will prevent disruptive surprises and give the new lecturer adequate time, often months, to prepare their materials.

5.2 Textbooks

The authors are in agreement that one reason the best-run courses are successful is because the course directors have found the proper materials to support the course objectives. In selecting a textbook and/or laboratory guide one must take into ac-

B.W. Newton et al., *How to Become an Effective Course Director*,
DOI 10.1007/978-0-387-84905-8_5, © Springer Science+Business Media, LLC 2009

count the content, depth of coverage, as well as the accuracy and timeliness of the material.

Textbooks must be ordered ahead of time so they will arrive before your course starts. This may require that a course director place the order with the local book store, or the administration may provide this service. Provide the textbook information to the academic dean so he/she can send it to the students. If you have a course website, post the textbook information. If this information is posted electronically, it is your responsibility to keep the information updated. It is the experience of the authors that some students, especially freshmen, purchase books far in advance of attending classes.

5.3 Syllabi

Syllabi seem to come in and out of vogue. Whether your syllabus is an outline or a proto-textbook, it must contain the important information that every student needs: course schedule, course rules, examination policies, etc. Course rules must be consistent with the academic policies established by your school, so it is important that the administration have an opportunity to review the syllabus before it is reproduced and distributed. Beginning directors should meet with the director of academic services to discuss the specifics of course syllabi production, including cost and the policies regarding printing and distribution.

5.4 The Classroom

In contemporary graduate and undergraduate education the classroom is still the venue where most course information is imparted to students, and it is important that teaching faculty be familiar with the proper use of classroom facilities. The classrooms must be scheduled for your course several months in advance, which necessitates having an accurate course schedule available at the time the classroom request is submitted.

5.5 Lecturing

Teaching faculty must be informed in a timely manner of the lectures they are to give, and when they are to be given. The timing of when lecture notes and handouts should be distributed to the class (i.e., several days before the lecture or during the lecture) is a point of negotiation with the faculty.

Lecturers should be made aware of the relationship of their lecture content to past and future course content. Additionally, teaching faculty should be familiar enough with the overall curriculum to see the importance of their lecture content in the context of the big picture. This allows them to understand the workload that the students are facing, allows them to make subtle but meaningful connections across course boundaries, and will allow them to avoid the appearance of being out of touch with reality. The easiest way to accomplish this is to have one or two meetings in advance of when the course begins to iron out the details of how the course will flow and how the teaching team will interact.

Each of your faculty should develop clearly defined learning objectives for each of their lectures. This will set the pace for the course, allow students to see what is expected, and allow other faculty to determine what has been covered and what is coming next. The learning objectives serve another, more important function: faculty should be instructed to prepare examination questions that fall in line with their learning objectives. This is very important because students expect to be tested on content that is identified in learning objectives, and not on trivia. Keeping exam questions on target by use of learning objectives is one of the most important things that you can do to ensure high course satisfaction.

Different faculty will use different teaching styles, and it is good advice to allow them to express themselves in a manner that makes them comfortable. Later in this book we will address the issue of teaching styles. Suffice it to say that the manner in which one lectures dramatically impacts student learning, and happy lecturers deliver effective lectures. Lecturing is a high profile activity and an uncomfortable lecturer will not perform up to full potential.

5.5.1 Adherence to core curriculum

One of the problems many course directors face is that their teaching faculty don't adhere to the core curriculum or course objectives. It is common for lecturers to teach students about their research area, possibly because they do not have the breadth of knowledge that allows them to comment on other topics that are being taught in the course. When teaching faculty veer off track they compromise the quality of your course but more importantly, they adversely affect the quality of student learning. Gentle redirection of the lecturer is the best approach if this begins to happen. Remind them that teaching core course concepts and principles is much more important than discussing research whose application to the subject may be decades in the future.

5.5.2 Team Teaching: Go With the Flow

Team teaching is not a new educational modality. Team teaching is the standard way of teaching at the medical school level, where several faculty members share a common responsibility to present course content. These faculty members usually step in and out of lecturing roles and it is important that the course director and the faculty are cognizant of the content that is being taught before and after each of their lectures. It is important that all individuals be willing to go with the flow and dovetail into the content and depth of the other lectures in team taught courses. If this simple rule is observed, a team-taught course can be as seamless as a course that is taught by a single individual.

Integrated curricula present a much more difficult challenge for team-taught courses. We will talk more about this later in the book.

5.6 Getting Information Across to Students

Most students have short attention spans. The reasons for this are myriad and complex, but the advent of advertisements and music videos with rapid-fire images, the use of news sound bites and cryptic text messaging are contributing factors to the shortened attention spans of students. The aforementioned, combined with an entertainment mentality acquired from passively watching TV, DVDs, and surfing the web has resulted in a generation of students who have turned the "Three R's" from "Reading, 'Riting, and 'Rithmatic" into "Recline, Relax and Receive" (quoted from E. Robert Burns, UAMS Department of Neurobiology and Developmental Sciences; personal communication). The cumulative result is that over the past several decades it has become increasingly evident that the use of traditional 50-minute didactic lectures has become a less effective method to transfer knowledge to students. As a result, class attendance often suffers and the students are left to learn the material on their own using a syllabus, canned notes, (i.e., student generated notes), and old examination questions, all of which may be of variable quality.

Decreased attention spans and the need for "entertainment" and engagement dictate that course directors investigate the use of a variety of active learning styles to address the current generation of students. Active learning can be achieved in the classroom through the use of problem-based learning (PBL), case-based learning, and team-based learning (TBL; Barrows 1994, Hark, Morrison 2000; Michaelsen, Knight, Fink 2004). Problem-based learning, where the learning objectives are student driven, is faculty intensive since small groups of 6-8 students are used with a faculty facilitator in charge of each small group. Case-based and TBL, wherein learning objectives are faculty driven, can take place in a large classroom setting where several faculty members can control the learning environment. Laboratory sessions should be, *ipso facto*, active learning environments that engage the student in applying their knowledge to solving clinically oriented problems.

5.7 Preparing Examinations

Preparing an examination that tests the course content at the appropriate level of rigor is the single most important thing that a course director does. The examination provides the student an opportunity to show you how he/she is doing, and students really do want to impress that upon you. Equally as important, the examination is your opportunity to set the standard for the course (and the college) to determine if the student has achieved the proper depth of knowledge and understanding to be certified by the institution. If you prepare the examination correctly, students will overlook minor course blemishes to praise the course. Do it incorrectly and students will take every opportunity to flail even the best-run course.

Given the grade-conscious nature of students it is safe to say that examinations are the single greatest currency for students. Appropriate examinations are those that test the core concepts and principles of your course in a manner that is sufficiently comprehensive to assess student knowledge, and is sufficiently discriminating to separate the excellent student from the average student, and the average student from the poor student. We have some suggestions for you to consider as you frame in your examinations.

It is the responsibility of the course director to set the parameters for testing and to ensure that their examinations meet those parameters. If an examination does not sample the full range of material presented within the unit to be tested, it will be mildly criticized by students. If an examination includes some material that is not in the unit that is being tested it will be harshly criticized. How could this possibly happen, you may wonder? In team-taught courses with multiple exam question writers, it is common for lecturers to be unfamiliar with the big organizational picture of the curriculum. They may submit questions that are off target and will defend them rigorously because they constitute finished pieces of work. They may not realize that the exam questions are inappropriate for the current examination. Yet, no matter how much effort is expended to ensure that coverage is fair, some questions may slip through the "filter" and they will be identified by students as unfair. It is the course director's responsibility to field claims of unfair questions.

It is the responsibility of the course director to set the degree of difficulty for examinations by selecting the appropriate mix of difficult and easy questions from the available question pool. If the exam is too easy, it will be mildly criticized by the top students in the class. If the examination is too difficult, all students in the class will criticize it. If some of the questions are easy and some blisteringly hard to "set the curve", students will think that the exam is unfair. Ideally the exam should contain questions that are reasonably uniform in difficulty. By far, setting the degree of exam difficulty is one of the most challenging tasks for a beginning director.

If the degree of difficulty exceeds expectation and the class average is too low, class morale will suffer and class performance will become an issue. One correction for this condition is to weed out the difficult questions and re-grade the exam to raise the class average. Although this solves the statistical problem, it calls into question the ability of the course director to set the degree of difficulty in the first place. The best students in the class will be very upset with this solution because you will be discarding questions that they have answered correctly and you will be reducing their "lead" on the rest of the class. The poorest students in the class will be very happy with this solution because they will receive a higher examination score, and perhaps in the process elevate their class ranking. All of the students in the class will assume that you are open to negotiations on which questions should be deleted from future exams and you will have set a dangerous precedent. *It is best to get it right on the front end rather than do damage control after the fact.*

It is good practice to have at least 10% more questions than needed for a given examination since, for various reasons, you may need to replace some of the questions that were submitted. Once the questions have been received hold a "triage session" during which the teaching faculty critique all questions. It is best if each faculty member has to answer each question without the correct answer marked;

this helps to ensure that fatally flawed questions and questions with multiple an-swers are identified.

It is the responsibility of the course director to assemble the examination ques-tions into a document that can be delivered to the students on examination day. Appropriate instructions must be included on the examination to permit students to take the exam without confusion. If the examination uses a paper and pencil format, give yourself enough lead-time to do a good job of assembling the booklet and getting it reproduced. A thorough proof reading (with the intensity you would normally reserve for page-proofs of scientific manuscripts) is an absolute essential step in exam preparation - don't assume that your question writers have done this for their own questions. The last step in the proof reading should be to check all headers, footers, titles and instructions. Then check the question numbers (to be sure that there are no duplications or deletions and that the total number of ques-tions is what was intended) and check every foil within every question to make sure that each has the appropriate range of lettering (A,B,C,D,E, etc.).

After the materials are copied, examine a sample to be certain that all pages are contained within the booklet and that all pages are legible. It is a good idea to print booklets in excess of what is required to cover any incomplete or damaged booklets discovered by students in the examination room. If you think that these obsessive-compulsive activities sound like a waste of time, then you have not at-tempted to administer an examination using a defective booklet.

If you are using computerized examinations it is likely your campus has devel-oped a system for inputting questions and getting things ready for exam-day. The support staff for these functions can assist you as you learn the specific tasks re-quired at your institution. If you do have administrative support, it is essential that they get the finalized exam prepared far enough in advance so that it can be loaded onto the computerized testing system without rushing at the last minute.

5.8 Administering Examinations

It is the responsibility of the course director to set the examination schedule and to administer examinations. Make certain that the exam schedule is clearly stated for students in their course materials and/or the class web site. Do your best to adhere to the published exam schedule. Any modifications in exam dates or exam times from the published schedule will always upset someone in the class and may cause absences. Be extremely conservative about moving exams just because the class requests it; invariably you will create more scheduling problems than you will solve if you begin to move exam dates around. The worst-case scenario is that you could create a situation in which the class thinks that they can set their own exam schedule through negotiation.

5.8.1 Exam Day

Examination day is usually a day of rest for the course director, which may lead to an elevated mood. Exam-days are very difficult, trying and stressful days for stu-dents, and this may suppress their mood. Be cognizant of this difference in attitude and try not to rub salt into the wounds of students while delivering the necessary announcements and starting instructions. Keep chatter and laughter to a minimum among the exam proctors/stewards, since this distracts the students.

Course directors and exam proctors/stewards must be prepared to deal with any type of emergency that may occur — and believe us when we say that they will occur! The authors have dealt with the following unexpected events: a student gets ill or faints during the examination, or has to be seated close to an exit because of gastrointestinal issues; the PA system or AV system does not work; the room where the exam is to be held is locked; the room is excessively hot or cold; an adjoining classroom is so noisy that it is disruptive to the students taking the exam; or the worst-case scenario, a fire alarm during an exam.

If an error is found in the test booklet just *before* the exam starts, then corrections can be announced to the entire class. If a student finds an error in a question, they will almost certainly bring it to your attention. If no one has yet finished the exam, then the correction can be announced to the entire class. However, if even a single student has finished and left the room you cannot correct the error for the remaining students, since those who have finished did not have the opportunity to hear the correction. (Remember the "fairness" issue discussed in section 4.1?) If an error is found, you may have to eliminate the question or accept an alternate answer during grading.

It is not good practice to answer student queries about a question during an examination for several reasons. First, by helping individual students, you unfairly give those students an advantage over others who did not ask questions. Second, when administering standardized exams, like those given by licensing agencies, e.g., the United States Medical Licensing Examination (USMLE) and the National Board of Medical Examiners® (NBME), students are not permitted to ask questions. By following these rules you help the students prepare for high-stakes licensing exams.

5.8.2 Absent or Tardy Students

Once the class is seated, perform a head count to make sure that all students are present. If there is an absence, initiate the plan for a make-up exam and gather the information needed to assure that there was a valid reason for the absence. It is wise to tell those students who commute to school to make allowances for potential traffic delays. Nothing rattles a student more that showing up at the last moment or worse, missing the start of an examination. Most colleges have protocols in place that specify the steps taken to deal with students who are absent or tardy on exam day. Check with your institution and follow their guidelines. If there is no college-wide policy, then determine what you want to do beforehand and stick to your protocol. It is good practice to tell the class if a student is absent and to warn them not to openly discuss the exam with other students until after the make-up exam has been given.

5.9 Student Appeals of Examination Questions

It is the course director's responsibility to investigate misgraded questions, and to make corrections that satisfy the class. Grading errors are unavoidable. In hand-graded practical exams it is extremely difficult to grade all questions consistently. In computer graded exam situations, the computer will not make a mistake but the course director or programmer (those who develop and input the exam key) may make errors. Try to take these little problems in stride and handle them graciously. The best approach is to consider how you would like to be treated if you were in the students' shoes and had a correct answer but did not receive credit. Never assume that the grading is perfect - anything can happen when exams are graded and any problem that may arise can be solved.

Make sure that the students and faculty are aware of the exam appeals policy. This information needs to be included in your course syllabus. Pre-appeal examination scores can be posted as soon as the scores are calculated. Be sure to inform the students that a revised scoring will be posted when the appeals process is completed.

Often appeals are written by students who are "fishing for points" because they did not know the correct answer. Although you may have to sort through this chaff, legitimate alternate answers will be found. Appeals often turn up problems in semantics that cause well-prepared students to select a wrong answer. It is for these reasons that the appeals process exists. If the course director responds in a professional and respectful manner, even to the inane appeals, it will result in enhanced student satisfaction. Furthermore, the course director's response to the appeal can correct a misconception by the student or class. If an alternate answer is found during the appeals process, remember to make the correction in your question bank so that a similar appeal does not occur if the question is used again.

To help the course directors, the UAMS College of Medicine has established a student examination question appeals committee (Hiatt, Menna, Petty, Hackler, et al. 2006). The purpose of this student driven committee is to screen the appeals as a means of eliminating redundant appeals, silly appeals and appeals that are not appropriately documented. Each of the appeals that the appeals committee forwards to the course director must be sufficiently referenced as a means of eliminating appeals that are, for want of a better word, "just griping." The course director then answers the appeal(s), in writing, explaining why it is valid or invalid and frequently corrects the misunderstanding. The appeals committee system works very well and serves three very important functions: it saves course directors' time, it requires students to synthesize understandable and referenced appeals, and it requires that students work together for a common purpose.

5.10 Grading issues

A course director must *never* assign examination or final course grades in a way that a student with a lower percentage score receives a better grade than a student with a higher percentage score. If a more qualified student is skipped over so that a less qualified student receives a better grade, a claim of unfairness will ensue and it will be well grounded.

In many instances, the break points for letter grades will shift slightly from the published breakpoints provided to the students. In team-taught courses, where the assignment of grade break points is via group consensus, many faculty members can be influenced by student personalities. One way to prevent student personalities from interfering with setting course grades is to remove the student names from the rank order list of final points earned in the course, making the fixing of the break points an unbiased process. Finally, it would not be fair to raise the bar for achieving a grade. Hopefully this is common sense.

5.11 Working with Faculty

Clearly the quality of one's teaching faculty will in large measure determine the quality of one's course. It is essential that you choose your teaching faculty wisely, based on their level of expertise and their ability to convey information to students in a clear, concise and interesting fashion.

5.11.1 Choosing Your Teaching Faculty

Choosing faculty is not as easy as it appears at first blush. There is a host of factors that determine the constitution of one's teaching faculty, including availability of content experts, availability of release time for clinical faculty, departmental traditions as to who teaches what, and when it is to be taught. Research schedules and administrative commitments can impact heavily on the availability of teaching faculty. It is also necessary to get the approval of the chairperson, as most chairpersons consider the deployment of faculty resources to be under their control.

It is not unusual to "inherit" a poor teacher or two, and in such cases you must do the best that you can. Beginning course directors have one advantage over veteran course directors in that they can start this relationship with a clean slate. One of the things that can be done is to offer poor teachers assistance in improving their teaching skills, assuming that they are amenable to doing so. Local workshops or training sessions would be a logical way to direct this individual to seek improvement. If sufficient improvement is not achieved, it may be necessary for you as a course director to recommend that a faculty member no longer teach in your course. In such cases you might have to appeal to your chairperson and/or your academic dean in charge of the overall curriculum.

Several decades ago it may have been possible for a new faculty member to have a fairly comprehensive knowledge of the material being taught in his/her field of expertise, and to be able to relate their knowledge to other basic science disciplines. However, since the 1980's the strong emphasis on the research training of graduate students (our future colleagues) has either dramatically decreased or entirely eliminated time for graduate students to develop a broad perspective of their discipline and to learn the basics of pedagogy. Graduate students are now being trained to have a very deep understanding of a very narrow research topic, unfortunately at the expense of learning how their research topic, or specialty graduate courses, fit into the big picture of their discipline. Combine this lack of breadth in knowledge with the exponential increase in scientific information and it becomes clear why we are losing quality teachers and potential future course directors. Furthermore, the exponential increase in scientific knowledge has become a hindrance to more seasoned faculty members who find it increasingly difficult to stay abreast in their field of study and find it necessary to abandon the classroom to remain competitive in seeking research funding.

Recently, at our institution, a group of students were anecdotally asked if they had a million dollars what would they change about their medical education. Without hesitation, the students responded that they wanted teachers who really wanted to be in front of the class and who were prepared to give sound, informative lectures. In short, they wanted teachers who cared about student learning. The students also stated that they didn't want faculty members assisting in laboratory sessions only because they were required to be there. The take-home message is clear: for greatest success, choose faculty who want to be teachers.

5.11.2 Training Your Teaching Faculty

Teaching faculty should always try to improve their pedagogical skills. New and seasoned teaching faculty members should be offered workshops on the principles of teaching, learning, student assessment and curriculum evaluation. The knowledge of the science and the art of teaching should be the cornerstone of your course and your teaching faculty should be made aware that you value education highly. As a course director you should be the clarion of a student-centered educational philosophy.

If students are provided with a lecture syllabus, then have your lecturers speak about their topics in the same order they are presented in the syllabus. It is frustrating to students when they have to flip through the syllabus to find what the instructor is talking about. Presenting information out of order should be evident to the speaker, since the whole class will be simultaneously flipping though their syllabus or scrolling on their computers. Request that lecturers review their presentations before they are in front of the students so they are familiar with the order of their slides. Knowing the order of the presentation will allow the speaker to smoothly transition from one slide to another and from topic to topic. Although this seems intuitive, for many instructors it isn't.

One common mistake that lecturers make is to speak longer than the time allotted. Exceeding allotted lecture time frustrates the students since their break time between successive lectures is shortened. Taking too much time is also unprofessional because it impedes the next speaker from being able to get the AV equipment ready for their presentation. It is also unprofessional to continue answering student questions, either in the classroom or in the hallway, after the next lecturer starts. Gently let the students know they should be attending the next lecture and that you will answer their questions after class hours.

Although an instructor may have presented a lecture numerous times, they must remember that this is the first time the students are hearing their lecture and therefore, they must stay upbeat and interested. Being "exciting" to the students can take many forms, some of which are good and others questionable. Providing an entertaining but information-light lecture will keep students happy but will not be in their best interest when they have to take comprehensive licensing examinations. Overly dramatic use of body language or frenetically moving about the stage or lecture room may keep the students awake, but will ultimately prove distracting. Above all, students desire to have course concepts and principles explained to them and, in the process be shown how the information is relevant to their profession. Therefore, a dynamic, organized lecturer fits their content into the "big picture" of the course objectives, usually in a seamless manner.

An effective educator uses a variety of teaching modalities to get their points across to their students. For example, a static PowerPoint slide presentation can be combined with a brief video or sound bite, or with overhead projection or white board technology. As a means of engaging student interest, the beginning of a lecture can be devoted to presenting a simple scenario/clinical case that can be solved by the students at the end of the lecture or at the end of a short series of lectures. Small group sessions or team-based learning exercises can be used in place of a portion of the lecture and lab components of the course. Audience response systems can be used during the lecture to ask questions of the class as a means of gauging their understanding of the material, as well as providing the students with an interactive learning environment.

Course directors should meet periodically with their teaching faculty to acquaint them with the goals and objectives of their course, preferably before the course starts. The course director should also address the temporal sequence of topics covered in their course and the reasons for the sequence, especially if their course is tightly interwoven into the fabric of a highly integrated multi-course curriculum. Teaching faculty need to see the big curricular picture as well as the small component parts.

A course director should meet with his/her teaching faculty to address the policies of their course as well as the policies of their respective institution as they relate to student education. It is very disconcerting when a faculty member fails to adhere to a policy since, as the course director, you must address the "fallout" of non-adherence. At times this can be very stressful, and even embarrassing depending on the nature of the non-adherence.

The course director should continuously evaluate the teaching effectiveness of his/her faculty, and those instructors who are poor teachers should be offered support and guidance to enhance the effectiveness of their teaching. This could be in the form of attending a teaching workshop or consultation with the school's office of educational development. Support of this nature helps to ensure a highly qualified teaching faculty. If the faculty development process fails, it is the responsibility of the course director to replace the instructor with one who is competent. It is recognized that such situations can be difficult at times but the major objective of the course director should be the presentation of the best course possible, and this at times mandates that he/she make some difficult decisions. The course director must have the support of the academic deans in making difficult teaching assignment changes when the teaching evaluation data indicate that change is required.

Finally, an effective teacher is not necessarily the most popular instructor. Students, in general, are not good discriminators of instructor effectiveness. Their teaching awards often unintentionally deteriorate into "popularity contests" instead of honoring an effective instructor. Peers can often tell who is an effective, dynamic instructor even though those defining characteristics are varied and subjective.

5.11.3 Training Your Teaching Faculty to Effectively Use PowerPoint

There may be some teaching faculty who are not familiar with the production of PowerPoint presentations. In such cases, it is your responsibility to offer to either personally train them or ask them to enroll in a PowerPoint training course.

PowerPoint presentations can be highly effective as a teaching and learning technology if used properly. Teaching faculty should be made aware that there are do's and don'ts in the use of PowerPoint presentations. The first "don't" is presenting too many slides. A good rule of thumb is 20-25 slides per 50 minute lecture. (And no cheating by making the print/pictures smaller so you can pack more information per slide!)

Another thing that should be avoided when using PowerPoint is the verbatim reading of slide material – nothing turns students off more than this! Most students feel that they would be better served if they stayed home and studied on their own. Students can read slides just as well as a lecturer. If your institution offers courses or workshops on the use of PowerPoint presentations it would be in the best interest of the students for you to ask your teaching faculty to attend.

5.11.4 Use of Audiovisual Equipment

The course director should know how to operate the projection equipment. Make sure you know how to dim room lights, adjust sound levels, etc. Have extra batteries on hand for laser pointers and cordless microphones. Anticipate problems with PowerPoint presentations and software compatibility. If there is any doubt about equipment compatibility, request that your lecturers bring their presentations on a CD as well as a flash drive. Always instruct them to install their presentation on the classroom computer before their lecture is scheduled to begin. Finally, know how to access the web if the lecturer is going to link to websites during their presentation.

5.12 Working with Students

As a course director you are expected to maintain a professional demeanor in all situations. Students can be frustrating at times, but you cannot lose your temper.

One situation that occasionally arises is that a course director reaches his/her breaking point and spouts off a tirade at the entire class. Usually it is one or two individuals who deserve the tirade, and the rest of the students just think that the course director is a jerk. Self-control is the key in these situations. It is possible to discipline an individual or a class in a respectful manner by letting them know you are disappointed with their performance or behavior and that you know they can do better. Most students don't know the difference between a professional vs. personal disagreement, so you have to use caution so that a student (or students) don't think you will hold a grudge against them for the remainder of the course.

Use common sense and not a "sledge hammer approach" when solving problems that can be dealt with in a less than draconian fashion. The same rule applies when dealing with faculty. Fortunately, most faculty members are mature enough to know that a professional difference of opinion should not overtly compromise their personal interactions with you.

If you make an error while lecturing or in administering your course, admit your error as soon as you find out about it and correct yourself. Students and faculty are generally very forgiving if you admit your mistakes. Never be afraid to say "I don't know", then offer to look it up and inform them at a later date. At times, students will volunteer the correct answer. Let this occur and thank them for teaching you something.

Finally, suppress the urge to provide historical perspective. Students *never, never* want to hear how rough you had it back when you went to school, or that the current course is "watered down" and is not as challenging as compared to previous years. Even if all of this is true, students live in the present, not the past. One thing you can do is use your own educational experiences to empathize with them, reassuring them that you know how difficult it is.

5.13 Working With the Administration

One concern of beginning course directors is - will I be able to work effectively with the academic administration of the school? Hopefully, the seasoned course director has already found that his/her academic dean(s) is/are helpful in the daily running of their course, and if not, has taken measures to improve this important interaction. New course directors are concerned about many issues; such as will the administration interfere with the way I want to manage the day-to-day running of my course? Not only that, but will my chairperson try to micromanage the daily running of the course, and in the process undermine my authority with the students and the students' perception that I am the course director? Will there be competing agendas from different levels of the administration, i.e., will the dean of the college present a different working philosophy than the academic dean? If so what does one do? Will I have the support that I need from the curriculum committee of the college? How will the curriculum be coordinated? Is the culture of the school supportive of education?

All of the aforementioned questions have answers that vary from school to school and from department to department within a school. The authors would like to offer advice regarding these questions based on our collective experiences.

Beginning course directors should meet with their chairperson soon after the decision has been made to become a course director. The chairperson should discuss in detail his/her philosophy for working with the course director so an excellent course can be presented to the students. The beginning course director should anticipate that the chairperson will at first monitor his/her actions closely, and once experience is gained a reduction in monitoring should occur. The beginning course director should nevertheless keep the chairperson updated as to problems that are occurring in the course.

The beginning course director should inquire of the former course director and the academic dean(s) as to the current nature of the curriculum and the curricular problems and issues currently being addressed. Importantly, the beginning course director should arrange to meet with the chairperson of the school's curriculum committee. This will reveal the short range and long range goals of the curriculum committee and how his/her course intercalates into the overall curriculum.

Never be afraid to approach your administrators for assistance. Their job is to facilitate your job as course director. When presenting the administration with an item for which you desire action, having a written account of what you need and offering some objective data to support your request is advisable. One way to request additional support for your course or faculty is to put these items into your course report. It is advisable to have a prioritized list of what you desire. When dealing with administrators, you have to realize that they are working under various rules, regulations, and budgets that constrain what they can do to fulfill certain requests. The administrators may fully agree with what you desire, but may not have the resources at hand to assist. Gentle persistence on your part will keep the issue in their minds so that when the resources become available they may be able to help. If there are ways that the administrators can be of better assistance, constructive comments are always welcome.

Being compliant to administration requests for information will keep you in favor. Many times, the administrators are in turn responding to requests for more information from outside sources. Keeping accurate records of your course is extremely helpful. In particular, a timely response in providing accurate information needed for accreditation site visits is essential.

5.14 The Chain of Command

The authors believe that a clearly defined chain of command is essential to the efficient implementation of a course. The purpose of the chain of command is to address student issues in a manner that is consistent, well defined, and fair in the eyes of the students.

In the chain of command, students should always go to the course director as the first step in addressing a course issue or problem. If after meeting with the course director the student feels that his/her problem was not addressed adequately, the student should go to the departmental/division chairperson for counseling and direction. If after meeting with the chairperson of the department/division the student still does not feel that he/she has received adequate redress, the student should go to the academic dean. If the academic dean is not able to settle the issue, the student has the option of seeking counsel with the dean of the college. Students should not jump links in the chain, and should this occur the student should be advised that he/she is to adhere to the chain of command.

This system has been used at the UAMS College of Medicine for over 20 years and has worked very well. During this time no student has sought the counsel of the dean of the medical school for redress of a course issue or problem. The authors highly recommend that course directors implement this very simple but highly effective chain of command.

6

Managing Courses with Laboratories

It is very important that the laboratory component of a course be managed in an efficient manner. As a course director you are the person who is responsible for the ordering of laboratory supplies and the reservation of laboratory space. It is also your responsibility to recruit teaching faculty to staff each laboratory session and to make sure that they are aware of the goals and objectives of each of the laboratory sessions. If teaching assistants are used in your laboratory sessions you must make sure that they are also aware of their role in each session.

6.1 Lab Hazards

Course directors must make sure that all teaching faculty and teaching assistants are aware of the potential hazards of each laboratory session and are knowledgeable as to the protocols for breaches in safety. Students should also be informed of potential hazards. When infectious agents are used in laboratory sessions it is essential that all laboratory participants be aware of the proper means of decontamination before leaving the laboratory sessions. In short, student safety is always an important consideration in lab sessions, and as a course director you must adhere to state and/or federal safety policies.

6.2 Clinical correlations

In the health professions, clinical cases can be presented at the start of the laboratory sessions with the aim of students gaining sufficient knowledge to answer questions pursuant to the clinical case at the end of the exercise. For example, in a neuroscience course "lesion hunts" can be performed wherein the students must identify the site of a lesion when symptoms and signs are provided and vice versa, i.e., give the deficits when a lesion is presented. Small group presentations and "chalk talks" are particularly effective in laboratory sessions. The faculty can probe the students for their level of understanding of major concepts and then spend time addressing areas of deficiency, either one-on-one or in small groups.

6.3 Faculty Work Load

The laboratory component adds contact hours to the workload of all faculty members involved. This can make faculty recruitment a difficult issue, as laboratory teaching generally flies under the radar of the administration and is often not re-

B.W. Newton et al., *How to Become an Effective Course Director*,
DOI 10.1007/978-0-387-84905-8_6, © Springer Science+Business Media, LLC 2009

warded. Laboratory sessions are generally interactive and the students can ask questions over any portion of the course (or even other courses). As such, faculty members must be competent in their discipline if they are assisting in laboratory sessions. Accordingly, faculty participating in laboratory sessions must be informed as to the objectives of each laboratory session and must be cognizant of how laboratory sessions fit into the course and the curriculum in general. If a faculty member does not have an adequate background, then other faculty members must be found who have expertise in that area of the course. This is not optimal since students often ask questions that pertain not only to the current laboratory session but also to former material that has been presented in the course.

One way for the course director to assist new faculty in the laboratory setting is to keep a list of the most commonly asked questions for each lab session. In each new class, most students will ask the same questions asked by previous classes. If the course director has such a list, and it is a good idea to keep one, then the list can be given to the new lab faculty member as a study aid. In essence, if they know the answers to these "standard questions", then they will have the wherewithal to answer over 90% of the questions they will be asked. Then with continued experience, they will learn how to answer the remaining 10%.

6.4 Tie Laboratory Content to Lectures

Laboratory sessions should be presented at a level commensurate with the educational level of your students. Laboratory sessions should be kept relatively simple, but at the same time bring didactic material into the realm of "discovery". If possible, laboratory sessions should be constructed such that students work together in the gathering of data and in problem solving. Not only is this a good learning experiencing for the students, it also allows you to observe students working with one another and affords you an opportunity to observe students' communication and leadership skills – not a small issue in education.

It is essential to tell students not to segregate the information they learn in the laboratory setting from information presented via other modalities. Although this seems like common sense, many students fail to use concepts learned in the classroom to answer lab exam questions and vice versa. For their part, teaching faculty must integrate laboratory and lecture information so that students view laboratory sessions as an integral part of the course. Once again, it is the responsibility of the course director to set the boundaries of what is expected of the students in the laboratory setting. This means that every laboratory session must have clearly defined learning objectives as well as study questions for student self-assessment.

The down side of laboratory sessions for students is that they are required to spend contact hours in a laboratory setting rather than using that time for independent study. The upside is that the concepts and information presented in lectures are usually put into practice in the laboratory component of the course. This permits students to perform a significant amount of their study under faculty supervision and, presumably, get questions answered and misconceptions corrected.

Therefore, if students take advantage of a well organized, effective laboratory experience, at least part of their heavy studying can be done before they return home.

6.5 The Gross Anatomy Course – It's Unique

If you are the director of a gross anatomy course you will encounter issues that are unique to the running of laboratory dissection sessions. In addition to the usual pressures of designing and implementing a major laboratory course, there are legal issues involved in the handling of human material. You are encouraged to become familiar with the laws of your state regarding the proper handling of human cadaver materials.

The authors have decided to address the general issues that are not bound by state laws to assist you in getting your gross anatomy lab running.

- An individual must be hired to embalm the cadavers and to keep the laboratory in good working order. It is highly recommended that this individual be a licensed mortician. The same individual will usually transport bodies to the University for embalming. A dedicated vehicle must be available for transport of cadavers.
- Campus housekeeping personnel must be instructed on how to keep the gross anatomy laboratory clean.
- Students must be told to be professional and treat cadavers with respect.
- Students must be trained to keep the cadavers in good condition. This is especially important if the gross course is taught for a lengthy period of time, i.e., over more than one semester, or spanning a winter, spring or fall break.
- Students must be taught to use the dissection equipment safely.
- The instructors have to be trained to triage injuries and determine if, in a rare instance, a cut is serious enough to warrant a trip to the emergency department for stitches. For the more frequent minor cuts, have on hand an ample supply of band-aids, alcohol swipes and/or iodine, and individual applicators of antibiotic ointment.
- Some institutions require all students to have a hepatitis B vaccination, and that is certainly good advice for students working with cadaver material.
- The gross anatomy course director has to comply with all regulations of the state funeral board. Meticulous records have to be kept on those who wish to donate their body but have not yet died, as well as maintaining records on when a cadaver was used, how it was used, and who received the remains.
- If the next of kin does not want the remains returned, you must make plans for proper burial.
- To prohibit voyeurism, allow only students and teaching faculty into the dissecting room. Photography should not be permitted without the approval of the course director and non-students should not be allowed in the gross anatomy lab "to see the cadavers". These are common sense guidelines to protect the privacy of the cadaver donors, and it is always easier on the course director if a strict policy is in place.

- You must make sure that the bookstore has the correct equipment for students to purchase. This includes dissection equipment, scalpel blades, gloves and lab coats.
- Due to space constraints, the entire class may not be able to take laboratory practical examinations at the same time. If not, a multiple exam system must be devised that maintains the security of the exam so that those students who finish the lab exam first do not share information with students who have not yet taken the exam.
- Students are generally very apprehensive about their first lab practical examination. The authors have found it useful to have the sophomore class set up a small practice practical examination before the first lab exam is given. With proper course director guidance, the practice practical can be representative of what the faculty assembles for a real examination. This small exercise is greatly appreciated by the freshmen and helps to relieve anxiety. (The same is true if you have a neuroscience/neuroanatomy wet lab.)
- If clinicians or residents are to assist in the gross anatomy lab they must be instructed on proper anatomical terminology and should be encouraged to read the dissection guide and attend the lecture(s) before they go into the gross lab. It has been the experience of the authors that clinicians may be "loose cannons" in the lab. They may distract the students from dissection time with anecdotes, or tell the students that what they are learning is not important, either because the clinician doesn't know the material or because the clinician considers it unimportant for his/her specialty. The biggest difficulty encountered when clinicians teach in gross lab is that they tend to use specialty-specific jargon and outdated terminology. The bottom line is to be very selective in whom you choose to assist in the lab.
- Many institutions have some type of remembrance ceremony at the end of the gross anatomy course to remember and honor those who were altruistic in the promotion of student learning. The course director is usually asked for input and help in assisting the students in organizing the ceremony. It is good advice to keep a file of the programs of these ceremonies so that a copy of previous programs may be given to the current student organizing committee.
- Gross dissection occurs in a team setting where 4-6 students dissect a single body. At times personality conflicts occur and a student may want to be moved to a different dissection table. It is the authors' opinion that the students must work through these conflicts on their own and stay together in their originally assigned group. Our logic is this; as medical professionals, students will not usually have a choice with whom they work. We feel this is an important first step for them to settle their differences and learn to work as a team. Only under the most extraordinary circumstances should a table assignment be changed.

7

Responding to Student Problems and Issues

The ability of your students to have a "safe learning environment" depends in large measure on the manner in which you treat them. It goes without saying that students should be treated with respect and dignity. They are adults and, with increasing frequency, many are married and have families. They should not be treated like children.

Always remember that common problems and issues are unique to the student even if you have encountered the same issues in the past. It is the first time your students are taking your course and they are constantly adjusting to the stress of learning. In the case of medical school, they are trying to adjust to a voluminous amount of material presented in a very short period of the time. Most schools have a support staff and facility for students, so it is important that a course director know what is available and how to direct students to get the help they may need. What follows is a brief guide of how to handle some of the more common problems encountered by students.

7.1 Social/Adaptation Problems

In this category of problems are things as diverse as separation anxiety from leaving family, loneliness in a new environment, the inability to find a compatible roommate, difficulties attempting to maintain a long-distance relationship, etc. Occasionally, a student arrives with poor social skills that alienate classmates and irritate faculty. These issues should be directed to the academic dean.

Often what students expect their educational experience to be like is incongruous with reality. From the authors' experience, this is especially true for medical school because past and current TV shows and movies have grossly distorted the reality of the day-to-day life of a medical student. Kay (1990) has called the student's introduction to educational reality as "traumatic de-idealization". This wake-up call to reality probably occurs in any profession that has been artificially glamorized by TV and movies, e.g., criminal law and forensic medicine.

7.2 Inadequate Preparation Before Entering Medical School

Not all undergraduate premedical programs are created equal. Some students may not have been subjected to the academic rigor that will prepare them to make the transition to medical school. In particular, many students haven't had a team-taught course where (at our institution) as many as 12 - 26 faculty members may

B.W. Newton et al., *How to Become an Effective Course Director*,
DOI 10.1007/978-0-387-84905-8_7, © Springer Science+Business Media, LLC 2009

deliver a course with 90-110 contact hours. Two solutions are offered: first, inter-
vention by the course director at an early stage of the course, before the damage is
irreversible; and second, get the student into a tutorial or educational skills program that is offered by your Office of Educational Development or like organization. The authors have found that the successful student has a strong work ethic,
and this ethic often overcomes an inadequate preparatory education.

7.3 Inadequate Study Skills

Students with inadequate study skills should seek professional help. You may be
able to make course-specific suggestions to help the student survive in your
course, but this type of student generally has academic problems in more than one
course. Get the student into a tutorial or educational skills program that is offered
by your Office of Educational Development or the equivalent.

Another option is to have the student talk to each of the course directors for
specific help in how to study a particular discipline. In this regard, it is good practice for all course directors to keep each other informed about who is in academic
jeopardy in their course during the entirety of each semester. If a student is in
academic jeopardy in more than one course then each course director can advise
the student to seek additional assistance from the Office of Educational Development.

7.4 Real or Perceived Lack of Study Time

Your Office of Educational Development, or like entity, usually has an expert who
is specialized in time management issues. If this is not available, one suggestion
that will work for many students is to have the student get a schedule organizer
and map out their entire week, blocking out study time, relaxation time, etc. Once
the schedule is set, they must have the discipline to adhere to it. This simple
method can work wonders for a student's academic performance.

7.5 Convincing Marginal Students to Accept Help

Students don't always want your help. One can only offer assistance – it is the
student's decision as to whether he/she accepts it. If the offer is made privately
and in a non-threatening manner, and if you demonstrate flexibility in scheduling
meeting times, you will generally get a positive response from a student. If a student rejects your offer for help, you have fulfilled your obligation as a course director. Regardless of the outcome, it is good practice to annotate your meetings
with students to document that assistance was offered.

7.6 Overcoming the "Grade-Oriented Persona"

Health professions students are an interesting group. They have taken undergraduate course after undergraduate course to score a grade that, in the aggregate,
was sufficient to get them into a terminal degree program. The mentality that "the
grade is the reason to take the course" is hard to deal with at the graduate or medi-

cal school level and takes a long time to "catabolize". This is why it is important to place your course material into the context of the students' career path and emphasize to students that what is being taught in your course is a tool they will use in their future career. Emphasize that your course is more than just an obstacle to overcome for a good grade. Your course presents valuable information that is a building block in the foundation of their professional careers.

7.7 Student Syndromes

Burns (2006) published a tongue-in-cheek article in which he described nine student "syndromes" that afflict the academic performance of beginning medical students. Some of these syndromes may apply to students in other disciplines. In brief, Burns' nine student syndromes are:

- "Six Chambered Heart" syndrome, a condition that results when students over analyze examination questions, which in turn, enhances the probability that they will choose the wrong answer.
- "Slip and Slide" syndrome occurs when students study for an exam in one course at the expense of keeping up with material being taught in other concurrent courses.
- "Oh Yeah!" syndrome occurs when students recognize the course content from their undergraduate days and do not study the material in a depth sufficient to perform well on medical school examinations.
- "Too Many Books" syndrome occurs when students feel that consulting many texts (or Web sites) will provide them information in a more concise manner or in more detail than the required course text(s) and/or course syllabus.
- "Irrelevant Material" is a syndrome where students fail to appreciate, or are not informed of, the relevance of what is taught.
- "Post-Genius" syndrome occurs when medical students are accustomed to achieving at a high level and may have never had a "C" grade in their academic career. Thus, earning a "C" or lower on an examination or as a final course grade is often devastating to their egos.
- "Alternate Student" is a syndrome in which a student who is accepted as an alternate to medical school feels that he/she is inferior to his/her classmates who were offered acceptances at an earlier date.
- "Recognize the Old Test Question" syndrome occurs when students who study from old exams jump at what they recognize as a correct answer on a current exam when, in actuality, the question stem has been changed so that the answer is now incorrect.
- "Memorize Everything – Understand Little" syndrome occurs in many students who are admitted to medical school. These students possess excellent memorization skills that allowed them to achieve at a high academic standing during undergraduate school, but often lack a realization of how information pertains to the medical profession. However, as undergraduates they were not challenged on examinations to critically think and apply the information that they memorized. Memorization sans true understanding of the material often results in poor medical school test performance.

8

Measuring Course Effectiveness

Measures of course effectiveness must be employed to continuously improve the quality of a course. There are many ways to measure course effectiveness and multiple approaches should be used.

8.1 Examinations

Teaching effectiveness is measured using examinations, quizzes and ultimately licensing examinations. We have already touched on examination content and how it affects student satisfaction. We would add here that, if possible, in-house examinations should emulate question and exam administration formats that are used by licensing examinations. For medical schools the National Board of Medical Examiners (NBME) Subject Examinations and the United States Medical Licensing Examination (USMLE) Step 1 Examination would be the pattern to follow. For other disciplines it would be a different format.

For medical schools, the use of multidisciplinary-examinations is in keeping with the nature of the USMLE Step 1 Examination, which is the first of three USMLE examinations leading to medical licensure in the United States. Given that all of the USMLE examinations are either partially or totally computer-based, the use of computer-based in-house examinations is a good method of simulating these high stakes examinations.

The use of NBME Subject Examinations provide, to some extent, an external outcome measure of course effectiveness (although the authors feel that such examinations are probably not as effective as in-house final examinations in most basic science courses). It is worth noting that the NBME now has customized Subject Examinations that afford course directors some latitude in choosing questions that are included on their examination. Customized Subject Examinations may prove to be a better comprehensive assessment of a student's overall knowledge of a course. The ultimate objective measure of the effectiveness of a basic science medical school course is how well students do on the USMLE Step 1 Examination. This is a comprehensive examination of all basic sciences courses taken at the end of the sophomore year.

B.W. Newton et al., *How to Become an Effective Course Director*,
DOI 10.1007/978-0-387-84905-8_8, © Springer Science+Business Media, LLC 2009

8.2 Student Evaluations of the Course

Student evaluations of a course are an essential metric of *perceived* course effectiveness. Although subjective in nature, student evaluations yield invaluable data that can lead to major course improvements.

Students should evaluate the course as soon as it ends. The instruments used may vary in breadth and in format. The typical evaluation tool would have a series of descriptors to which students respond by agreeing or disagreeing. The important thing is to create a series of descriptors that frame the evaluation process and gather input on potential trouble spots. Computer-based evaluation instruments afford easy student access and allow quick feedback to course directors. Paper and pencil evaluations are also commonly used and are equally effective.

Whether the format is computer based or paper and pencil, student comments should be encouraged. Student comments can be very informative and supplement the numerical data that is obtained from evaluation descriptors. The collection of student comments allows students to touch upon topics that are not included in the assessment descriptors. Even more important, it gives students a chance to raise issues and vent frustration. This helps students develop a sense of empowerment and a feeling of contributing to the betterment of the course.

A follow-up report to the students, following their course evaluation, can yield a big boost in student satisfaction. If changes are made in a course in response to student evaluations, the students should be informed of those changes. It is also appropriate to inform students in the next academic year of changes in the course made in response to previous student evaluations. Informing the incoming class of changes sends a clear message that student opinions have credence - this is a huge point often neglected by course directors. Not informing the new class of student-generated course changes makes the changes transparent to the students. In addition, informing each incoming class of changes generates better student compliance with the evaluation procedure.

In curricula employing sections or blocks that are multidisciplinary, each block should be evaluated as a means of determining areas of perceived deficiency. Frequent student evaluations of a course allow for real time modifications in a course, and the process demonstrates to students that a course director and the teaching faculty are reactive to their views in a real time setting.

Student course evaluations should be reviewed by the curriculum committee or its designated subcommittee and by the academic deans of the medical school. Based on student evaluations of a course, the curriculum committee should make recommendations or mandates to the course director for changes in the course.

8.3 Faculty Review of Student Evaluations

Course directors should review the outcome of the student evaluations of the course with their teaching faculty. This is best done by meeting with key teaching faculty to review student evaluations and addressing each of the major concerns of the students. This process gives the faculty an opportunity to get the information straight from the students, rather than filtered through the course director. It is not uncommon for faculty to perceive the outcome of the evaluation differently from the way the course director sees it and the ensuing discussions are always productive for all involved. Although the intent of the course evaluation review is to make corrective changes in the course, it can also be quite rewarding for faculty to see what the students think is being done well and to receive some praise from the students.

8.4 Student Evaluations of Faculty

It is advisable to have the teaching faculty evaluated after they give their last lecture in the course. This is recommended for two reasons. First, it will give the faculty some formative feedback directly from the consumer. Second, it will develop a paper trail of faculty performance that can be utilized in promotion and tenure decisions.

If your course is team-taught it is a good idea to include on the evaluation form a photo of the instructor along with the titles of their lectures to avoid confusion. If possible, have an impartial third party collect the evaluation data and keep it until the course is over, then distribute it to the faculty members. This assures the students that their comments will not negatively influence their grade and removes the course director from the direct line of student to faculty assessment.

8.5 Faculty Evaluations by the Course Director

It is the view of the authors that the course director should evaluate each untenured faculty member every year prior to their application for promotion and tenure. He or she should be given a written report regarding their teaching performance. Be sure to mention strengths as well as candid comments regarding points needing improvement.

It is also valuable to have faculty outside of your department evaluate untenured faculty members who are lecturing in your course. Their unbiased input can be instructive to the lecturer and is useful in the promotion and tenure process. Unannounced peer evaluations will reveal the average performance of a lecturer. This prevents a lecturer from putting in an extraordinary effort on a day when he/she knows they will be peer evaluated. However, there is nothing wrong in informing a faculty member at the beginning of the term that a certain number of their lectures will be peer evaluated.

Tenured faculty should also be periodically student and peer evaluated in order to document their continued effectiveness as an instructor. Once every three years is usually sufficient.

The course director should keep all faculty evaluation data for the entire term of his/her directorship. Faculty members do not always attach significance to their annual teaching evaluation data and are likely to misplace the documents. Be assured that as faculty are preparing documents in support of their promotion and/or tenure, they will be contacting you for their student and peer evaluation data even though you have provided it to them on a yearly basis. Do not expect the academic dean to keep faculty evaluation data – it is his/her responsibility to keep course evaluation data, not faculty evaluation data.

8.6 External Course Evaluations

Some schools, including the UAMS College of Medicine, conduct external course/clerkship evaluations. An external evaluation is a thorough course review conducted by a visiting faculty member who is a subject matter expert, and usually a course director in the discipline which is being evaluated. At UAMS College of Medicine, this process is carried out by the curriculum committee working with course directors and the chairpersons of their respective departments and occurs in each course or clerkship once every seven years.

The protocol for the external course evaluation at UAMS is as follows:

- The course director submits to the curriculum committee the names of three potential evaluators along with a *curriculum vitae* for each.
- The External Evaluation Subcommittee of the College of Medicine curriculum committee reviews the credentials of the potential evaluators and makes a recommendation to the full curriculum committee as to their choice of an evaluator.
- At the monthly meeting of the curriculum committee, the full committee reviews the *curriculum vitae* of each potential evaluator and a vote-based decision is made as to the selection of the evaluator.
- The External Evaluation Subcommittee then notifies the selected evaluator as well as those not chosen.
- The course director submits a course self-study to the evaluator (containing standard information along with anything else the course director would like to submit) and works with the External Evaluation Subcommittee in arranging the evaluation process. The External Evaluation Subcommittee also supplies the evaluator with the protocol of the external review process and the level of detail expected in the evaluator's report.
- On completion of the site visit (usually two days), the evaluator submits his/her report to the chairperson of the External Evaluation Subcommittee who then distributes the report to the members of the subcommittee. The evaluator then receives his/her honorarium.
- The report is distributed to the course director, the departmental chairperson, the dean of the college, the academic dean, and the members of the curriculum committee.
- The subcommittee then drafts a letter to the course director in response to the evaluation informing him/her of the perceived strengths and weaknesses of the course.
- Once the subcommittee letter is approved by vote of the curriculum committee it is sent to the course director and his/her chairperson.
- The course director is given a deadline to respond, in writing, to the contents of the evaluation letter.

. If your school does not utilize external course evaluations, it would be advisable to suggest to your administration that they do. The use of the external evaluation process at the UAMS College of Medicine has proven very effective in addressing chronic course problems by emphasizing the strengths and weaknesses of courses. It has provided a stimulus to the administration to identify resources for courses when it was previously thought that problems were not significant.

8.7 Other Evaluations

Another area of evaluation the authors would like to address is the use of student forums. At the UAMS College of Medicine, and other medical schools as well, student forums are held to review courses and curricula (Harden, 2000). It is imperative that the person(s) facilitating the forums be well qualified. It is suggested that the selection of the students to sit on the forum panel represent a cross section of grades allocated in the course – this helps to ensure a diversity of opinions. The questions asked at student forums should be general in nature, thus allowing the faculty moderator latitude to pursue issues as they arise during the dialogue. It is imperative that the faculty moderator be someone that does not teach or assess the students, i.e., a person who is readily identifiable by the students as neutral. The faculty moderator should advise the students at the outset of the forum that he/she is neutral and that comments will not be specifically attributable to anyone on the panel. This will assure the students that they are in a safe zone and can talk freely. The use of student forums has been used at the UAMS College of Medicine with great success and has revealed issues not identified in the general student or faculty evaluations of a course. This has proven to be a very helpful in the improvement of our curriculum.

9

Documentation

Course directors have the responsibility of keeping records on all types of course activities. These range from keeping grade records to recording all of the subtle changes that may occur in the course curriculum. Importantly, it is always wise to record who teaches what and when. Knowing this at the outset can make a big difference in recording accurate information as it happens. A few suggestions are given below.

9.1 Archive the Grade Book

When the course concludes, it is the course director's job to assemble the final grade book and report the grades to the registrar. Archive a copy of the grade book in your own files and also in the departmental files. This creates a paper trail that is necessary to justify the function of the departmentally-based course and the performance of students who will be receiving a diploma from the institution. You will be surprised at how frequently you will review the old grade books to retrieve information when writing letters of recommendation for students.

9.2 Keep a Tracking Log of Course Changes

Course directors should keep a log of the changes that are made in their course from year-to-year. Doing so affords the course director and administration a record of the changes that were made and why they were made.

One reason to track course changes is that no one has the time to "reinvent the wheel" if a course evaluation reveals that a particular course of action was not effective in improving the course. Reverting to a previous, more effective variation of the course is simple if the changes are clearly recorded.

A second reason to keep a tracking log of course changes becomes obvious when a school undergoes an accreditation site visit. Having a log of course changes justifies to some degree why a course is taught in the current fashion.

Finally, the tracking log makes it easy to communicate course changes to the curriculum committee, to other course directors teaching in the same year of the curriculum, and to the academic dean.

B.W. Newton et al., *How to Become an Effective Course Director*,
DOI 10.1007/978-0-387-84905-8_9, © Springer Science+Business Media, LLC 2009

9.3 Write a Course Report

Writing a course report is an excellent idea for course documentation, even if your institution does not require one. A course report serves as a thumbnail sketch of the course and can be an invaluable tool to use when making decisions for staffing and course modifications. It will also provide a valuable record of course history when a new course director takes over.

At a minimum the course report should include the number of contact hours and a listing of what type of contact hours they were, e.g., lecture, lab, small group sessions, examinations, etc. The course report should also contain a listing of who participated in your course and the nature of their contributions. The course grade distribution, class grade point average and, if applicable, class performance on the national standardized subject examination should be included in tabular form. Including several years of historical data in this table provides an efficient year-to-year comparison of course effectiveness. Include student evaluation data (sans written comments), the texts that were used in your course and any substantial changes that you made in the course from the previous year (i.e., include the tracking log of course changes).

In addition to being an historical archive, the course report is also a planning document. A list of strengths and weaknesses provides talking points that can be utilized when engaging faculty or administration in discussions of course outcomes. List any future changes you desire to make in your course and the reasons for the changes. Finally, a list of needs that the administration can address is a great way to conclude the report.

As part of the documentation process, course directors should submit the course report to the academic deans and to the curriculum committee or, if one exists, the evaluation subcommittee of the curriculum committee. The departmental chairperson should review the course report before it is sent to the aforementioned committee or subcommittee.

Once the academic deans have received the course report, the course director and his/her respective chairperson should meet with the academic deans to discuss the report. The substance of the meeting should then be reported to the curriculum committee and/or the evaluation subcommittee of the curriculum committee. This is the process that is used at the UAMS College of Medicine and it works very well. There have been many times that a course director received assistance that would not have been rendered in a timely manner had such a meeting not occurred.

9.4 Archive Instructor Evaluation Outcomes

As mentioned previously, faculty members often do not file their teaching evaluation materials carefully. Keep a copy of the instructor evaluations in a file that is organized by year. This will give you the documents needed when you are asked to provide instructor evaluation materials several years after the fact.

10

Impediments

We literally spent hours discussing the impediments that course directors face while conducting our IAMSE workshops on how to become an effective course director. It is noteworthy that most of the concerns of the beginning course directors were the same as those of the veteran course directors, although the relative importance placed on each impediment varied. While it is impossible to list all of the impediments that were discussed at the workshops, we have addressed herein the ones that we think are the most pertinent in becoming an effective course director. Importantly, we offer advice on how to deal with each of the listed impediments.

10.1 Time Constraints

Beginning course directors often wonder if they will have sufficient time to run a course given that they have many other responsibilities including research, graduate student teaching, seeing patients (if a clinical course director), academic and non-academic committee assignments and involvement in professional societies. Veteran course directors probably have addressed the issue of time constraints. The first suggestion we have for beginning course directors is to meet with their departmental chairperson and discuss the chairperson's plan for how to distribute your time. This will identify the time constraints, gauge the limit of the chairperson's support and allow you to focus on course administration tasks.

The second step for a beginning course director should be to meet with the previous course director to discuss the issue of time management. Veteran course directors know how to do everything that the course requires in an efficient manner and their suggestions can help you conserve as much of your time as possible.

Another suggestion is for all course directors to delegate some of the course responsibilities to responsible faculty members. Once the course responsibilities have been delegated, the course director should meet regularly with his/her assistants to keep abreast of their progress and to address problems with the common goal of presenting the best course possible. The course director should not micromanage the work of his/her assistants.

Finally, if your school has an educational intranet site, beginning and veteran course directors should be shown this site and told to make frequent use of it. Course directors need to familiarize themselves with the faculty, student and course director handbooks. Utilizing these resources will save time.

B.W. Newton et al., *How to Become an Effective Course Director*,
DOI 10.1007/978-0-387-84905-8_10, © Springer Science+Business Media, LLC 2009

10.2 Lack of Faculty Incentives

Course directors are always concerned about whether or not they will have a sufficient cadre of good instructors to teach in their course. What is the genesis of this concern? Some basic science chairpersons tell their faculty that promotion and tenure are based on research productivity and will certainly not be granted based on exemplary teaching or for directing an excellent course! Whether stated or intimated this message is heard loud and clear by master teachers who either stop teaching or greatly reduce the time that they teach (Hafferty, Franks, 1994). This is a loss for students as well as for the faculty member for whom teaching may be a calling. Yet, this phenomenon continues to erode the very fabric of education and not much is being done to correct the problem.

How does a course director overcome the impediments in establishing a cadre of excellent teaching faculty? Suffice it to say that the solution exceeds the authority of the course director. Indeed, the school and the curriculum committee have the ultimate responsibility for presenting a top-notch curriculum to their students. As such, addressing these impediments is not the sole responsibility of the course director.

If a course director is encountering problems in the acquisition of good teaching faculty he/she should discuss the issue with his/her chairperson, assuming that the course is run out of a department or division. It is the chairperson's responsibility to assist the course director in obtaining highly qualified teaching faculty. Second, the course director should inform the academic dean responsible for the overall curriculum that faculty resources are needed. This can be initiated through the "Needs" section of the course report and lead to a working meeting with the administration and the chairperson that is aimed at obtaining sufficient qualified and committed teaching faculty. Once again, the school has the ultimate responsibility for providing sufficient numbers of competent teaching faculty.

10.3 Micromanagement

Course directors frequently express dismay over the top-down administrative tactics often employed in the management of the curriculum. For example, there have been situations in which the dean of the college mandates a change from one major teaching modality to another – often engendering indifference among faculty members. Given that the dean is the CEO of the school there is little that faculty can do other than accept the charge.

It is important to note that the purpose of academic administration is to facilitate education. It is the function of the academic administration to ensure that regulations and policies are adhered to by course directors and teaching faculty and, at the same time, afford the faculty the academic freedom that ensures a dynamic and vibrant curriculum. There are times however when the academic administration oversteps the line and infringes on academic freedom. At such times, course directors need to address the situation with their chairpersons, who in turn should address the concerns with the administration. Generally, this is sufficient to remedy the problem.

It must be remembered that frequently, the administration is responding to guidelines from a higher authority. For example in a medical school, educational standards are constantly being modified and new ones implemented in response to the directives of the Liaison Committee on Medical Education. Some of these standards require modifications in educational modalities and this can be difficult for faculty members accustomed to teaching the "old way". During such times the course directors often feel that the administration is micromanaging their course. Accordingly, the administration needs to meet with course directors to better facilitate the process of change by informing the course directors that the administration is simply complying with regulations imposed by professional licensing/accrediting agencies.

10.4 Breaking the Rules

Sometimes, faculty members don't accept direction. It is safe to say that in just about every school there is a faculty member who, for whatever reason(s), chooses not to follow course and/or college policies. One often sees this phenomenon in senior faculty members who began their teaching careers at a time when policies and curriculum governance were different and teaching faculty had more autonomy in making "on the spot" policy decisions.

Course directors must insist that all teaching faculty adhere to course policies and the academic policies of their school. If written policies are not followed to the letter, a school may find itself embroiled in a lawsuit wherein a student is claiming unfair treatment. As discussed previously (section 4.1), the issue of fairness is extremely important to students, especially with regard to policies that govern student assessment.

How should a course director address an incident of non-compliance with institutional policy? The first thing that must be done is to discuss the problem with the offending faculty member. The faculty member should not be accused of wrong doing – just informed that he/she is in non-compliance with a policy. It is suggested that the course director meet alone with the faculty member with a copy of the policy in hand. The conversation should be one of mutual respect with the course director explaining to the faculty member the nature of the policy and why it is in place. The consequences of non-adherence to the policy should be emphasized with the discussion centered on what non-adherence does to the course. During the meeting, the course director should ask the faculty member if he/she has other things to discuss about the course, since there may be other issues which the course director is unaware that engendered the breach in policy. The key to the successful closure on such problems is to be non-accusatory and to let the faculty member know that he/she is a valuable asset to your course. At the close of the meeting the course director should ask the faculty member for their future cooperation with regard to the policy issue being discussed.

On occasion some faculty members will be less that cooperative when discussing their violation of a course policy or an institutional policy. Some faculty may say that the policy is silly and of no consequence; others will say that the policy prevents them from exercising their academic freedom as teachers. Others may harp on accreditation or curriculum committee "mandates". Still others will say that they just don't like the policy and they are not going to comply. In such situations it is important for the course director to ratchet up the pressure – effective course directors do not allow "rogue faculty members" to undermine the quality of their course! If that does not solve the problem, then the next step entails addressing the issue with the chairperson of the department or division responsible for the course. In the case of a course that is not run out of a department or division, the issue should be taken to the academic dean. The chairperson or the academic dean must meet with the faculty member and explain to him/her the nature of the policy to which they are non-adherent and request that they become compliant. Many times this is sufficient to resolve the problem. When it is not, the chairperson or the academic dean should inform the faculty member that he/she is no longer allowed to teach students. This may sound heavy handed - but it is justified.

10.5 Personality Conflicts

Many of the conflicts between faculty members have little to do with a substantive issue and more to do with personality differences. Personality conflicts are a way of academic life and course directors must be aware of this simple law of nature – people sometimes just don't get along!

How does a course director address incidents wherein faculty personality issues are creating problems in the effective administration of their course? First of all, the course director should not choose a position as to who is right and who is wrong – you have to live with both of these people. It is best to appeal to the faculty members' sense of professionalism. The course director should advise each of the faculty members that their differences, and the resulting conflict, are adversely affecting the course and as a course director one cannot allow this to happen. The course director should request that the faculty members cease from further obstreperous behavior. Above all, they must be informed that their issues should never erupt in front of the students. If at this point arbitration is unsuccessful, the course director should move up the chain of command and address the problem with his/her departmental or division chairperson. Generally this is sufficient to bring about closure. If not, the final solution is for course director to replace both of the "combatants" with other instructors.

10.6 Faculty Who Won't Accept Direction

Another impediment that concerns course directors is whether or not they will be able to work effectively with their teaching faculty. It is a fact that not all faculty members accept direction. This potential impediment is difficult to address since teaching faculty are generally quite diverse in their views as to how students should be taught.

A new faculty member who is joining a teaching team is usually not a problem. He/she is joining an established situation, they are trying to fit in and they are usually willing to accept direction. The course director should offer guidance as to the overall goals and objectives of the course as well as the rules and policies that govern the administration of the course. This information is usually gratefully accepted by a new faculty member and gets them started on the right foot.

At the other end of the spectrum is the senior faculty member who has been around while the current course was being developed, likes it the way it is (or not) and who has a pretty good idea that he/she can run this whole course by himself/herself. This type of faculty member may not accept guidance willingly. He/she may insist on giving the same lectures that he/she has given for many years – often without updating their lecture materials.

Senior faculty members who refuse to update didactic sessions should be encouraged to consider new findings relative to their lectures and be informed as to how the limited/antiquated scope of the material they teach is negatively impacting other material being taught in your course/curriculum. The faculty member should be lauded for the many years of teaching they have provided for students and should be encouraged to continue to teach. However, if the faculty member refuses to update his/her lecture material, he/she should be removed from the teaching roster of your course. Once again, it is the responsibility of the course director to ensure that the best possible course is presented to students. This means that no faculty member, either new or senior, is autonomous in their teaching – they are responsible to the course director and to the sponsoring department or division.

Where does the departmental or division chairperson fit into this equation? In some schools the departmental or division chairperson makes all of the decisions as to who teaches in the course they are sponsoring for the school. This can complicate the situation, for the chairperson may not agree with the course director as to who does or does not teach in "their" course. It is therefore incumbent on the course director to provide the chairperson with data as to the teaching ineffectiveness of those faculty members who fall short of effective teaching. Such evidence includes student evaluation data, analysis of the effectiveness of the faculty member's teaching performance and if possible, peer evaluation information. Unfortunately, even if convincing evidence for poor teaching is offered, the chairperson at times will decide to allow the faculty member to continue teaching. At this point the course director must do the best that he/she can do under such circumstances. It is an understatement to say that such occurrences severely undermine the effectiveness of a course director.

The importance of the chairperson in the chain of command of course administration cannot be overestimated. The chairperson can either be a positive or a negative force for a course. Occasionally, the chairperson will actually run the course behind "closed doors" and make the major decisions for the course director, often without knowledge of the students' perspective on a given issue. This puts the new and the veteran course director in a position where he/she is simply a "puppet" for the chairperson. Students are perceptive, and soon learn they may not be able to count on their course director to address their concerns. This leaves the students conflicted since they have the expectation that a course director makes the major decisions regarding course issues, and does so with the knowledge of their concerns and with their well being in mind.

10.7 Curricular Conflicts – Range Wars!

Course directors must work with other course directors and faculty members teaching in other courses. This is always important but becomes an even more important issue when a course director is administering a course in an integrated curriculum wherein all of the courses for a given academic year are running concurrently. In an integrated curriculum the decisions made for a course not only impact that course but all of the other courses. In such a curriculum it is simply essential that course directors be cooperative and see the curricular picture beyond their course boundaries. Generally, in an integrated curriculum the oversight is by one of the academic deans of the college and his/her immediate administrative assistant. This assistant is generally responsible for clerical issues attendant the co-ordination of the courses, scheduling the didactic components, and assisting with the assessment process and curriculum evaluation.

Unfortunately, it is inevitable that conflicts will arise between course directors in an integrated curriculum regarding who teaches certain material, when certain lecture topics are taught, how many questions per didactic session should be on the examination, the nature of the examination questions, policies that govern the administration of the integrated curriculum, etc. It is the experience of the authors that course directors generally address these conflicts on their own and to the satisfaction of the other course directors administering the integrated curriculum. If an issue is unresolved, it is brought before all of the respective course directors and the academic dean for discussion. It is important from the onset of course integration that all course directors recognize that the integrated curriculum is administered in a manner that best serves the educational needs of the student and not a given course or faculty member – in other words the students *must* be the focus of the process. After discussion, the final decision(s) that are made are generally the result of compromise, but on occasion must be made by the academic dean.

10.8 Is My Course Budget Adequate?

Where does the money come from to run a course? Generally, the money comes from the dean of the college and is allocated during budget negotiations with departmental and division chairpersons.

Let's address what the cost issues are in running a course. The course director needs funds for the copying of course syllabi and lecture PowerPoint files, as well as laboratory equipment and supplies if the course has a laboratory teaching component. Occasionally the course director will require funds for the purchase or development of teaching modules for their students. If the course is synchronously or asynchronously delivered to sites off-campus, then funds to obtain and use educational equipment must be made available. In some institutions, media/academic services departments charge for the use of specialized equipment.

Funds will also be required for the updating or purchase of software programs that are required for the course, including the purchasing of software licenses.

In some courses, the department or division responsible for the course provides a course coordinator to assist in the development and presentation of the course, and in the synthesis and administration of course examinations. For those courses that do not have a course coordinator the responsible department or division should provide the course director with secretarial assistance. Also included in the budget should be travel funds for the course director to attend at least one education meeting per year. Unfortunately, the reality of appropriating budgets often makes attendance at educational meetings the first line item to be cut.

One also has to consider the cost of new equipment and the maintenance of old equipment such as computers, computer projectors, laser pointers, etc. It is the view of the authors that these costs should be borne by the university. It is the responsibility of the university to provide an adequate and modern teaching infrastructure for their teaching faculty; such costs should not be borne by a department or division.

The authors feel that educational costs for a course should be an earmarked budgetary line item, and should be detailed as to the reason(s) for the allocation. It is not unusual for money allocated for course support not to be a line item, but to be buried in the overall departmental budget. This is less than ideal, since the money allocated for running a course can be spent on other things – leaving the course director to make do with what remains. In short, the course director should make sure that he/she has a reasonable budget for their course and if not, should consult with their chairperson for needed funds. If funds are not available, the course director should request that the chairperson seek additional funding for running the course. Finally, obtaining funding for a course is the responsibility of the administration and not the direct responsibility of the course director.

11

Pitfalls

In our many years of experience, we have come to realize that there are a number of pitfalls that await the unsuspecting course director. The topics that are included in this chapter were lessons learned the hard way and we thought it would be a good idea to share them with you.

11.1 Never Let Them See You Sweat

The role of a course director has been outlined in great detail in the previous chapters. One thing that has not yet been stated is the effect that the demeanor of the course director has on the overall operation of the course. The course director is the ultimate course leader. Faculty members look to the course director for guidance, organization and occasionally, courage under fire. Students look to the course director for guidance and decisions. Everyone looks to the course director for his/her reaction when the going gets tough. If there is one trait that you should strive to develop, it is the ability to remain calm and placid on the surface while experiencing gut wrenching anger, irritation or frustration. Everyone expects the course director to control the situation - you can't afford to go ballistic!

11.2 It's *Always* the First Time

The information in your course is novel material for each new class that is admitted. Current students will ask the same questions and encounter the same issues and problems as your former students. It is your job as a course director to assuage student concerns and address their frustrations and questions as if it was the very first time that you have addressed such issues. Simply said, you must remain fresh in the eyes of the students and show an omnipresent concern for their learning and for their general wellbeing.

11.3 Don't Take It Personally

You are the lightening rod! If anything, *anything* goes wrong with any aspect of the course you will be struck. Expect to receive what you may consider unduly harsh criticism or complaints. You simply cannot please everyone all the time and a disgruntled individual taking aim at you is generally not representative of how the other students or your faculty feel about you. Remember, practically none of the faculty, and absolutely none of the students have ever been in your shoes and

B.W. Newton et al., *How to Become an Effective Course Director*,
DOI 10.1007/978-0-387-84905-8_11, © Springer Science+Business Media, LLC 2009

they may not be fully aware of the day-to-day responsibilities of a course director. Many will not have a complete understanding of the rules and regulations under which you must direct your course. Fellow faculty may not realize that you have to fit your course content into the overall educational objectives established by your curriculum committee. Gently explaining your situation to those that are complaining often resolves the problem. Be certain not to come across as whining about your responsibilities and the rules under which you must direct your course (after all, being a course director was your idea - or maybe not).

Although you will get complaints about a variety of issues from students, it is best to look for trends. If one or two students complain about a certain issue it probably doesn't warrant much concern on your part. But if many students come to your office, or you have dozens of responses on the course evaluation that address the same issue, the students have identified a problem that needs to be examined more fully and corrected. Even worse, if the same issue keeps appearing on course evaluations year-after-year, then action must be taken. This cannot be allowed to continue.

One way to head off unexpected student issues is to have periodic meetings with the class officers. If your budget permits, provide lunch – if you feed them they will come! The ambience of the meeting should be such that the students can discuss issues and concerns in an informal setting. Ask the class officers how they perceive the progress of the course and find out if they are aware of any negative undercurrents arising from their class. Showing concern and responding to student issues will be deeply appreciated by the class as a whole. Addressing small issues promptly will make it easier for you to make potentially unpopular, but educationally sound, decisions concerning larger issues at a later time.

Finally, be confident of your abilities and continue to improve your course on a yearly basis. Seek advice from the other course directors on how best to respond to student criticism. Remember, every year that you direct a course you gain experience and it becomes easier to respond to student issues and criticisms.

11.4 Overloading the Students

It is important to remember that students are taking other courses concurrently with yours. In order to afford students the time for studying your course material, your faculty members need to keep the reading and homework assignments reasonable. Providing concise reading assignments accompanied by tightly organized learning objectives promotes efficient use of students' study time. The use of several focused study questions relating to each lecture topic will help the students confirm they are on track and studying efficiently

Regarding medical schools, it is the opinion of the authors that there is far too much detail being taught in most courses. Reducing course content to that which is clinically applicable will enhance student critical thinking skills, since this will leave time for the lecturer to explain how the content is used. This approach, combined with focused reading assignments wherein students can learn the details, will result in improved learning and higher examination scores. Licensing agencies and the Association of American Medical Colleges (AAMC) recognize this and are pushing for medical schools to teach critical thinking and lifelong learning skills.

11.5 Learn to Say "No"

The word "no" is the most effective management tool you have in your arsenal. It is the key to implementing your important course direction decisions. If you say yes to all requests and demands, then who needs you? Learn to say no in a way that does not inflame the situation. It is definitely not a good idea to just say no without an explanation of the reasons or extenuating circumstances. A brief explanation of the reasons that caused you to say no is the best approach.

11.6 Avoiding Ennui

After several years as a course director, boredom and burnout may begin to set in. You may feel that you have "been there, done that" a number of times, but it is important to continue to pay the same attention to detail that you did during the first few years of your directorship. Always remember that organization is an essential key element students look for in a course and in a course director. The small day-to-day details necessary to keep the course running smoothly will start to quickly unravel if you think the course can operate on autopilot.

As stated previously, a course director must realize that each new class of students has not experienced your course. Your new students will ask the same questions about the same difficult course concepts as your former students. Although you have heard the same questions numerous times in the past, you must be on guard not to respond in a curt or condescending fashion; respond in an understanding and patient manner. Being patient towards each new class of students must be part of the orientation you, as a course director, give to your teaching faculty who lecture in your course year after year.

11.7 Preparation of Poor Examinations

As mentioned previously, examination outcomes are the student's bread and butter. Nothing is more frustrating to students than to take an examination that is poorly designed. Poorly designed examinations have questions that do not reflect the major learning objectives of the course, contain too many very difficult questions, contain ambiguous questions, and/or contain typographical errors. Course directors should allow a great deal of time for the proper preparation of examinations, and this means making sure that teaching faculty submit their examination questions in a timely manner. To repeat: the submitted questions must be clear, adhere to the prescribed format, and reflect the course learning objectives.

11.8 Intentional Inflexibility

Even though flexibility was listed at the bottom of the student rank-order list (see Table 3), as a course director it is still necessary for you to put yourself "into the shoes of your students" so you can respond to student issues and concerns in a compassionate and empathic fashion. However, there are some areas in which flexibility is not such a good thing. There are several of these scenarios that seem to recur, and we have touched briefly on each one herein.

11.8.1 Social Obligations Interfere with Education

Every student will have social situations that, at times, can have a detrimental impact on their ability to concentrate on their studies. Common examples include breaking up with a significant other, divorce, a request to participate in a wedding, family vacation plans, presenting research data at a meeting, and serious illness or death of a close family member. As the course director, students will come to you with these issues and ask for an excused absence from coursework or an examination. Each situation has to be considered individually. In general, the authors have not granted requests to travel to weddings or to accommodate family or student vacation plans if the request for leave overlaps an examination date. We feel it is not appropriate to grant such requests because academic calendars are usually provided to students far in advance (3-5 months) of when a course starts, and although some lecture times and topics may change, the dates of examinations are usually not altered. At the UAMS College of Medicine, the academic dean sends a list of all examination dates to students during the summer break between academic years and advises students not to plan weddings, vacations, etc., on those dates.

Some situations are unavoidable, such as the death of a close family member. We recommend that you define in advance how far out into the family tree you would grant a request to attend a funeral and then apply that policy if the need arises. It is a good idea to check with the academic dean, since a policy may already be on the books to deal with this occurrence. Cultural sensitivity must come into play, since what is considered a close family member can differ from culture to culture.

11.8.2 Grade Remorse

It is common for students to ask for a special project to improve their grade, usually after it is too late to do so by the conventional methods of studying and taking scheduled examinations. Be aware that if you assign an extra credit paper or allow a person to take an exam a second time, and you do not allow the entire class the same option, you are showing favoritism.

As discussed previously, many students entering medical school have never received a "C" grade in any of their prior undergraduate courses, and when they obtain their first "C", "D" or "F" on an examination they are emotionally devastated. Such students will seek reassurance from you that they have the ability to pass your course. Reassure them other students have been in the same situation and have worked diligently enough to pass the course. Be prepared for the student to say they know more than what their exam score indicates. It is your responsibility to probe the student for their knowledge of basic concepts. If a weakness is found either correct the weakness yourself, or have the student receive tutoring from the faculty member who gave the lecture(s). It is good practice to remind the student of their responsibility to learn the material in a timely fashion since future course material builds on former concepts.

11.8.3 Students Want to Redesign the Course Schedule

Even though you have a perfectly good examination schedule posted well in advance, students may still request changes in the timing of examinations. Often these requests will compromise their ability to learn in your course or other concurrent courses. In such cases, you must not make concessions that can hinder some or all of the students from being able to achieve at their highest level of performance. Inform the students that course policies and the timing of examinations have been carefully considered and that your experience has shown the suggested change would be detrimental to student learning. (Nevertheless, be forewarned that no matter when you schedule examinations, there will always be student discontent.) If that does not settle the issue, an effective way to defuse a request for an exam schedule change is to say you will change the examination schedule if 100% of the class agrees to the change and signs a petition to that effect. It has been the experience of the authors that getting 100% of the class to agree on anything is virtually impossible. Yet, giving the students the opportunity to vote on a change lets them feel they have some degree of control over their education and that you are willing to consider their request.

If you still have an overwhelming urge to change an exam date, be aware of two facts. First, if you try to move an exam up to an earlier date, someone will show up at the originally scheduled time and place to take the exam. Second, any perturbations in the schedule have a ripple effect that could adversely affect some aspect of the course that is scheduled days or weeks in the future.

At times, student requests are reasonable. Concessions that do not compromise learning issues or contradict college policies will go a long way towards promoting good student-faculty relations. An example may be a request to shift a lecture topic from a Friday afternoon to a Monday morning. If the lecturer is willing to do so, and the time slot and lecture hall is available, then a change like this would not compromise student learning. Minor changes in day-to-day scheduling are not a big deal — as long as *all* the students are made aware of the change in a timely fashion.

11.8.4 Students Want to Determine the Course Content

Nowadays, most students expect PowerPoint presentations during a lecture and a list of approved web sites for additional information. If you do not provide informative, accurate web sites that contain approved information they will surf the web and find information that may not be accurate. This information can draw students away from the stated learning objectives and cause them to spend time on material that you would consider trivial – reducing class performance on your examinations. Student dissatisfaction will be the result.

As a course director, it is your responsibility to decide what the content of the course should be and to define its boundaries. It is your responsibility to decide which learning modalities are most appropriate for the students and direct them to one or two good web sites or computer-assisted learning modules so the students can experience self-directed learning exercises. It has always been the case that students will follow direction, but if they receive none they will strike out on their own. Regardless or your direction, they may strike out on their own anyway (see section 7.7, "Too Many Books").

11.9 You Can't Do It All — Delegate

You may be tempted to think you have the ability (and the time) to do everything needed to run your course, and that you can do it better than anyone else. This could be true, but unless your administration grants you 75-100% time and effort to be a course director (doubtful), you may be setting yourself up for rapid burn out and potential failure.

One of the most important things you can do to help reduce your workload is to delegate some of the responsibility for running your course. For example, if there is a laboratory component to your course, appoint a laboratory director to assume responsibility for the day-to-day running of laboratory activities. Other tasks that can be delegated are coordinating the production of the course syllabus, updating the course website, and coordinating the scheduling of your course with the campus academic support office. Unless the delegate is doing a disastrous job, live with your decision and let that person organize and carry out their assignment in his/her own fashion. Of course, it is important that you support each person to whom you have assigned a task and make your expectations very clear.

One caveat: As course director the ultimate responsibility falls on you to make sure your course is organized. As stated previously, a good philosophy to follow is, "If something goes wrong with the course, the course director takes responsibility; if something goes right, then the faculty members get the credit."

11.10 Know Your Audience

It is a sad commentary that lecturers often do not understand the level of their audience. It is often the tendency of basic science faculty to deliver course material to health professions students at a graduate student level rather than at a level appropriate for medical students. It is a common mistake for clinical faculty to present to the freshman medical class a Grand Rounds lecture designed for residents. Simply put, too much material is presented at a level beyond the comprehension of the audience. Considering the large amount of material that health professions students are required to master in multiple courses over a very short period of time, the most effective use of your time and the student's time is to adequately cover major course concepts rather than talking about extraneous material that is of no relevance to the students' future profession.

The phenomenon of teaching at a graduate student level rather than at an undergraduate health professions level is often observed with new teaching faculty.

New faculty members are often unaware of the level that information should be pitched to health professions students. In such cases, it is the responsibility of the course director to meet with the faculty member and advise them to cut back on the amount of information that is given in their lectures. This is generally a painful process, for the new faculty member often doesn't know what should be omitted. Accordingly, the faculty member should be advised of the core goals and objectives of their lectures and, just as important, the goals and objectives of your course. Additionally, the new teaching faculty member should be informed of how their teaching sessions fit into the overall curriculum. Unfortunately, there may be times you will have to dictate to the faculty member what to teach – this is no easy task!

11.11 Make It Relevant

Making your course material relevant to the discipline for which the students are studying is very important for any profession. In the realm of the health professions, putting clinical relevance into your course makes it applicable to the future profession of your students and this makes the course more interesting. Examination questions using clinical stems enable students to know how they are doing in the application of key concepts and principles you wish for them to master. Students can now access details via PDA's or approved websites, so teaching clinical concepts vs. having students memorize a plethora of details will promote student satisfaction. Educating the student on how to apply information they have learned to novel situations (i.e., critical thinking) is one of your most important and most challenging tasks. Equally important in teaching health professions students is training them to use electronic databases and to apply the information they retrieve to the "big picture". Encouraging the application of material that students find in databases is a means of promoting lifelong learning.

11.12 Not Staying Current

After a number of years as a course director the temptation arises to let your course "run itself" (see section 11.6). Ultimately this may lead to dated course content or not paying attention to the day-to-day details of course organization.

Within the basic sciences, most disciplines have a knowledge base that expands on an almost daily basis. The course director and teaching faculty have to keep the course current and, at the same time, not overburden students with too much additional detail. It is good practice that when new information is added to a course something in the course needs to be either deleted or reduced in depth of coverage. Remember, in medical school the students are being trained as general undifferentiated medical practitioners – not graduate students. For example, when one of us was the Neuroscience Course Director (BWN), only three or four additional sentences were added each year to the entire course syllabus even though neuroscience is a rapidly advancing field. Only clinically applicable information was added.

11.13 Don't Be Afraid to Ask for Help – "You Are Not Alone"

Take advantage of the fact that there are many faculty members who have been in course director's shoes. Most faculty members are willing to help you deal with student issues and concerns the first few years you run your course – the time when you will need the most assistance. The academic deans and pertinent chairpersons should be able to assist you with various issues and the university policies concerning students and courses. If available, be sure you receive, *read, and understand*, a copy of the "Course Director's Handbook" and the handbook on campus policies.

Educationally-based Web chat lines and Web cast seminar sessions can be used to obtain advice from educators around the world. Attending meetings that address pedagogy in your discipline, and meetings devoted to more general educational concepts, can be very informative and invigorating. Organizations solely dedicated to education, such as the International Association of Medical Science Educators (IAMSE), are invaluable sources of support and information. You'll soon learn that the educational problems you are having with your course, your students and your faculty are universal in nature.

12

Professionalism

In response to an open-ended question asking which student issues make their job difficult, veteran course directors listed unprofessional behavior as the number one problem. Unprofessional behavior cited by the course directors included:

- Cheating
- Rudeness
- No concern for others, i.e., a cut throat attitude
- Lack of respect for authority
- Walking out on lecturers
- Habitual latecomers who disrupt the class as they enter and get seated
- Constantly pushing the boundaries of institutional rules and regulations

12.1 Bent Twigs

Changes in societal norms, with the consequential development of situational-ethics and the general lack of a moral compass, are contributing factors to unprofessional behavior. This phenomenon has eroded professional and societal responsibility wherein students feel the need to give back to society what society has provided to them - the privilege of attending medical or graduate school. Our current students have matured in an age where they engage in forms of electronic entertainment, on-line gaming and text-messaging. Such pastimes can isolate the person, and in the process stunt the development of socially acceptable face-to-face verbal and non-verbal (body language) communication skills, idealism and innate empathy, all of which continue to decline during medical education (Branch 2000; Griffith, Wilson 2001; Newton, Barber, Clardy, Cleveland, O'Sullivan, 2008). It is within this culture that faculty, course directors and the administration must instill a sense of professional responsibility and humanism into their students.

As a course director you will find that each class of students has a small cadre who never seem to be on time, never get their assignments done, miss mandatory class sessions, don't adhere to various campus policies, or have an outright disrespect for authority figures or their classmates. All of the aforementioned constitute unprofessional behavior.

Studies by Papadakis and colleagues (2004; 2005) have clearly shown that there is a strong association between students exhibiting unprofessional behavior

B.W. Newton et al., *How to Become an Effective Course Director*,
DOI 10.1007/978-0-387-84905-8_12, © Springer Science+Business Media, LLC 2009

in medical school and subsequent disciplinary action by state medical boards when they become practicing physicians. This lends credence to the proverb "As the twig is bent, so is the tree inclined". We mention this to impart the idea that tolerance of student "quirks" may not be the best approach to developing a responsible graduate.

12.2 Current Faculty and Administration Views Regarding Professionalism

Professionalism is rapidly becoming a very important issue in medical education as public approval of physicians declines (American Board of Internal Medicine, 1995; Pellegrino, 2002; Inui, 2003). The LCME and the Advisory Council on Graduate Medical Education (ACGME) have issued accreditation standards to ensure that medical students and residents receive a formal curriculum on professionalism. These standards also state that medical students and residents must receive evaluations of their professionalism (LCME, 2008).

In response to these standards, medical schools are beginning to establish codes of professional conduct for medical students in the classroom, the clinic and the hospital. In many medical schools, professionalism is now one of several competencies in which medical students must show proficiency as a requirement for graduation.

The establishment of student honor codes and professionalism as a competency that must be met for graduation are supported by the administrators of medical schools and are generally implemented as a function of the curriculum committee with the backing of the faculty. The proverbial "writing is on the wall" – health professions students are being closely scrutinized with respect to professionalism and this will continue!

12.3 The Course Director's Role

Many schools have clear guidelines that define what is and what is not acceptable professional student behavior and it is the responsibility of course directors to know these tenets of student professionalism. In the health professions, basic science course directors will frequently opine that, as basic science faculty, they are not "equipped" to address unprofessional student behavior because only persons in the same profession (i.e., an MD or RN, etc.) are capable of doing so. This is definitely not the case, for course directors should not only assess student knowledge but also student professionalism. Even though the basic scientists in your course may not consider themselves to be role models for health professions students, their behavior is being watched by students.

Unprofessional student behavior that is allowed to go unchecked has a way of being perpetuated in a student's clinical years of education and beyond. When this occurs the patient is the one who suffers. It is the responsibility of all health professions faculty to ensure that their graduates have developed the knowledge and professional demeanor to continue their education as residents and beyond.

12.4 The Student Honor Code

Many colleges have established an honor code that is enforced by a student honor council. One of the more common honor code infractions that student honor code committees encounter is cheating on examinations. Students, faculty and examination proctors can inform the student honor council that an honor code infraction has occurred. As a course director, you should follow the rules established by your college as to who can report an honor code violation.

12.5 Off-Campus Breaches of Professional Conduct

Some violations of professional conduct occur outside the classroom. Egregious offenses involving the city, state or federal authorities must be brought to the attention of the administration. Generally, a course director is not involved in the resulting deliberation as to what should be done with the offending student.

12.6 Improving Medical Student Professionalism

Many medical schools have taken advantage of a program established by the Arnold P. Gold Foundation called the "White Coat Ceremony." During this ceremony faculty members and student leaders speak on various aspects of professionalism with the ultimate goal of communicating to the beginning medical student that professionalism is the cornerstone of the empathic and competent physician. After the charges to the class have been made, the students recite a code of professional conduct, and then sign a book pledging their allegiance to the code. They are then donned with their short white lab coats to wear as a reminder that they must exhibit professional behavior as students and future health care professionals. This ceremony, given at the start of medical school, helps inculcate the concept that professionalism begins immediately. Other disciplines may have similar ceremonies promoting professionalism.

12.7 Recognizing Student Professionalism

Excellent professional student behavior should be acknowledged. Faculty, administrators and students should be given the opportunity to vote on awards given to students who exhibit exemplary professional behavior. In turn, students, faculty and the administration should vote on awards to give to faculty whom they feel exhibit outstanding professional behavior, humanism and compassion. These faculty members should be highly sought after as role models for giving lectures or rounding with students on the wards.

Health professions schools should have courses and extra-curricular events devoted to ethics, humanism, and professionalism. The concepts and scenarios discussed in these courses should be reinforced by the faculty who teach in other courses and by the administration. One method of tracking the development of professional, humanistic, and ethical behavior is to have students periodically write in a portfolio about their experiences as they progress through their education. Anonymous excerpts can be drawn from these portfolios and presented to the class for discussion.

13

Administrative Support

There are two sides to every story. Up to this point we have presented information derived from our combined experiences as course directors. We have emphasized the role of the course director and the expectations that a course director should have for students and the administration.

Two of us have many years of experience on the "dark side" – i.e., as administrators in positions that are referred to in this book as "academic dean". In this chapter, we want to present the administration's view of the interactions that occur between course directors and administrators.

13.1 The View from the Ivory Tower

Let's address what the administration expects of course directors as a means of seeing this essential interaction from the "eyes of the administration." At a meeting, academic deans were asked to rank order a list of course director attributes relative to their perceived importance. The results are shown in descending order of frequency in Table 5.

TABLE 5
Attributes Academic Deans Feel Course Directors Should Possess
(N=75)

Enthusiasm for teaching
Communication skills
Organizational skills
Knowledge of how the course fits into overall curriculum
Knowledge of assessment methods
Time management skills

When comparing these results to the results presented in Chapter 3, one startling outcome emerges: administrators ranked enthusiasm as the number one course director attribute, while course directors and students ranked it fifth. Apparently, administrators don't want to deal with grumpy course directors.

Administrators ranked communication skills second in importance. Veteran course directors and students also ranked it second (see tables 2 and 3, respectively), while beginning course directors ranked it fourth. It is not clear whether

B.W. Newton et al., *How to Become an Effective Course Director*,
DOI 10.1007/978-0-387-84905-8_13, © Springer Science+Business Media, LLC 2009

the administrators were ranking the ability of the course directors to communicate with administrators or their ability to communicate with students.

Administrators ranked organizational skills third (as did the students) while verteran course directors ranked it second (see Table 2) and beginning course directors ranked organizational skills first (see Table 1). It is refreshing to see that administrators and students have similar attitudes concerning the organizational skills of course directors. However, those in the hot seat ranked organizational skills higher.

The working relationship of course directors and academic administrators is impacted by the pedagogical needs of the students, the teaching faculty, departmental chairpersons, the dean of the college and the academic governance body of the school (generally the curriculum committee). Nevertheless, there are essential duties that are carried out by course directors in concert with academic deans that are essential to the effective running of the curriculum.

Course directors are generally responsible for the production of course schedules. It is the responsibility of a course director to work with other course directors in the production of a composite course schedule free of scheduling conflicts. It is the responsibility of the administration to interact with administrators from other colleges at their institution to avoid cross-college conflicts. If conflicts arise that cannot be resolved by the course director, then the academic dean who is responsible for the production of course schedules must work with the directors to resolve the conflict. It is always best to keep in mind that the objective is not necessarily to schedule what is best for a given course but to schedule what is best for an effective flow of didactic information.

The administration is responsible for making sure that the course directors are informed of any conflicts and, if required, that changes are made in the schedule. In universities with medical schools, administrators from other colleges often view the college of medicine as the "eight-hundred pound gorilla" in the room. While course directors might assume that the college of medicine should always prevail when scheduling conflicts occur among other colleges – this is not necessarily so. Course directors need to understand that the academic deans of the various colleges also have to work in a spirit of mutual cooperation, and that the resolution of scheduling conflicts has to be a give-and-take process.

13.2 Curricular Improvement

The administration expects course directors to work with the curriculum committee in effecting curricular change when the faculty of the school has approved change. Course directors, like other faculty members, are recalcitrant to change the curriculum especially when the changes involve their course. Multiple forces including inherent resistance to change engender this recalcitrance.

Often academic deans hear the common refrain, "if it ain't broke don't fix it"! Although the curriculum may not be broken, it is often in need of routine maintenance to enhance student learning. Academic administrators also hear course directors complain that they are too busy to work on curricular modifications, and that they hardly have enough time to carry out their responsibilities as a course director and perform essential research or clinical duties. The veracity of this statement is unquestionable for course directors have many other responsibilities.

In many medical schools, the aforementioned problem is exacerbated by the academic and monetary reward systems wherein basic science faculty receive significant incentive pay increases for receiving NIH grants, while they receive little or no incentive for directing a major medical school course. The end result is the marginalization of medical education. This marginalization of medical education is discussed in depth by Abrahamson in his wonderful book titled "Essays on Medical Education" (Abrahamson, 1996). Clearly, it is the responsibility of the dean of the college and other academic administrators to reverse this very negative trend, and this is beginning to occur in some U.S. medical schools. In short, the administration must reward course directors for their essential contributions to medical education, and add merit incentives for those directors who administer an excellent course. Excellence can be evidenced by student evaluations, by NBME Subject Examination scores and by USMLE Step 1 scores in areas covered by their course; the latter two metrics being more of an objective outcome measure of the quality of a course.

The administration expects course directors to work with one another in a spirit of cooperation and in a professional manner. In those curricula that are integrated and use integrated examinations, it is incumbent on the course directors to work together in the synthesis of effective examinations. Course directors should meet to review all of the examination questions from all of the courses represented on an examination to make sure that redundant questions are addressed as well as to check the clarity and correctness of the questions. Speaking from experience, such meetings at times can be contentious, especially when one course director submits an examination question that involves subject matter being tested in another course at the same time. These occurrences at times require arbitration from a third party, generally the academic dean. The "proprietary rights" of course subject material should always be respected since this represents an "ownership" of subject material.

13.3 Keep the Faculty "In the Loop"

The administration expects course directors to regularly communicate with their respective faculty members the status of their course, student issues, student grades, and problems they are having with the course. Communication with their teaching faculty will help to ensure a smoother running course and one wherein all course faculty members have a say in the presentation of the course.

13.4 Informing Course Directors of Current Educational Trends

It is *sine qua non* that the academic leadership of a school stays well informed of the current trends in education. Attending and participating at national and international meetings on education, and consummate reading of the current educational literature facilitates this process. Information gained from attending meetings and reading current education literature must be communicated to the curriculum committee membership and to course directors.

Not all new trends in education "fit" all schools. Ideas brought back from educational meetings and presented by academic administrators simply may not work in a given local setting. New trends in education should be discussed by the curriculum committee and by the course directors to assess the likelihood that implementation would enhance student learning. Course directors should be given license to attempt new educational modalities. Furthermore, it is suggested that a course director who wants to try a new educational modality consider running a pilot program to ascertain whether the modality has educational merit. The running of a pilot program is dependent on the support of the curriculum committee and a funding source. In any case, it is important that the pilot project be evaluated by both students and faculty to see if it is worth continuing and, possibly, expanding. In all cases, the results of the pilot study and/or expanded program should be submitted for publication.

To catalyze the process of curricular improvement, the administration should provide funds for course directors to attend education meetings. The course directors in turn should be expected to present summaries of the most interesting sessions that they attended to their faculty, the other course directors, the curriculum committee, the academic deans, and to the Office of Educational Development, or similar body. Such communication stirs the pot of mutual interest and sends a clear message to course directors and faculty alike that education is important at their institution.

13.5 Provide Monetary Support to Run Your Course

It is the responsibility of the dean of the college to ensure that all courses and clerkships receive sufficient funding to carryout their educational responsibilities. As stated previously it is the responsibility of the course director working in concert with his/her chairperson to make sure that the dean of the college is fully aware of the cost of running their course. The costs that are incurred in the running of a basic science course include the cost of supplies and equipment for laboratory exercises, the cost of maintenance contracts on equipment not normally funded by the school, and the cost of printing course syllabi and other handout materials for the students. Additional costs include software that will enhance the teaching effectiveness of the faculty.

Many basic science courses are now using computer-based examinations and the use of such examinations requires the expenditure of funds for software licenses and the funds to hire an administrative assistant (examination coordinator) to utilize the software in the preparation of examinations. In the case of integrated curricula, the use of computer-based examinations is warranted by the number of courses that are contributing questions to an examination and the frequent need for graphics in the synthesis of high quality examination questions. We have found at the UAMS College of Medicine that an examination coordinator is essential in the preparation of our integrated freshman and sophomore examinations.

Although generally not essential, it is appropriate for schools to provide funds for the recruitment of outside speakers as a means of exposing students to eminent scientist-educators. In the case of medical schools it is sometimes possible to obtain funding through the state medical society. It is the responsibility of the course director to make the necessary arrangements to recruit the very best speakers for his/her course – preferably speakers who are not only noted for their expertise in their field of interest but also ones who have the ability to teach effectively – no small task! It does students little good to listen to one of the "giants" in a given field of science or medicine and be overwhelmed by the volume or complexity of the material being presented. Although there are risks in having an outside speaker present to your class, it can also be very interesting and at times educationally exciting for the students to listen to those who work on the cutting edge of their field of interest. Once again, it is the course director's responsibility to ensure, as much as possible, that outside speakers are good teachers as well as being distinguished in their field of expertise.

13.6 Provide Support to Obtain Quality Teaching Faculty

The course director and his/her immediate superior (usually a departmental chairperson) are responsible for ensuring that the course has sufficient, highly qualified faculty to deliver the course. Such support also includes the time for faculty members to teach and, in some instances, monetary incentives for being a course director. Generally, the responsible chairperson meets with the dean of the college during budget negotiations to hammer out the budgeting of the course. The course director should communicate effectively with his/her respective chairperson to make sure that the chairperson is fully aware of the needs of the course.

As mentioned previously, medical education is being marginalized by the demands for clinical and research revenues. These demands leave little time for many faculty members to teach. This problem is exacerbated by the lack of incentives for educational activities. Educational activities are generally not rewarded, or if they are, not at a level that is equitable with rewards given for the acquisition of research grants and the accruement of clinical dollars. Given this scenario, it is imperative that course directors make their need for more teaching faculty known to their academic deans. Generally, medical school deans are aware that education is one of the key functions of a medical school and are eager to provide a quality program provided that sufficient funds are available.

13.7 Educational Support

It is the responsibility of the administration to identify funding for faculty development. Most faculty members have had no formal training in the theories of learning, i.e., the meta-cognitive components of learning. As such, faculty members many times teach in the manner that they were taught – and not infrequently this leads to ineffective instruction. Course directors should be aware of this problem and strongly suggest to their teaching faculty that they attend workshops on teaching and learning and attempt to put into practice what they were taught. This is especially important in the case of new faculty members with little if any teaching experience.

The dean of the college and the other academic administrators should be aware that faculty members well versed in the meta-cognitive components of learning and in tune with the latest didactic approaches in education are better equipped to teach students. Having educationally-informed faculty members enhances the likelihood that students will do well on their board examinations. Therefore, it is essential that the administration recognizes the need for faculty to receive the release time necessary to improve their teaching skills.

It is essential that course directors and the curriculum committee have the resources that are needed for the planning, implementation, and evaluation of the curriculum. The Office of Educational Development should serve as a beacon of educational scholarship, and their faculty should work closely with course directors to assist them in the presentation and publication of innovative educational modalities, innovative assessment methods, and methods of course and teaching faculty evaluation.

Educational scholarship is the final product of educational innovation and should not only be supported monetarily but also in the promotion and tenure process. Accordingly, it is incumbent on the dean of the college to establish an environment in which educational scholarship is nurtured and rewarded.

13.8 Providing and Maintaining Audio Visual Support

It is the responsibility of the administration to adequately support the office responsible for providing audio and visual (AV) aid to instructors. Many schools are part of a campus containing multiple colleges where an AV office/department generally serves as a central support service. In such cases, the funding for AV services comes from the chancellor or someone of equal rank. It is the responsibility of the school administration to make sure that the school has the required AV support that is both modern and reliable. Funding for AV support should be a budgetary line item with built in increases to offset the negative impact of inflation in terms of procurement and maintenance of AV equipment. The development of new technologies must also be taken into consideration as electronic equipment and software can quickly become outdated.

It is the responsibility of course directors to make their AV needs known to the administration. It is the responsibility of the academic dean responsible for course scheduling to ensure that he/she knows the AV needs of his/her course directors, and to ensure that those needs are met. The quality of educational instruction is dependent not only on the abilities of the instructor, but also on the quality of the AV support that is rendered.

13.9 Production of Course Director's Manuals

It is the responsibility of the administration to provide the course director with a manual of educational rules and policies that are concisely written and understandable. All schools should supply their course directors with a course director's manual. Having a manual is a boon to course directors for it helps make their job a great deal easier in addressing the day-to-day problems of running an effective course. Once again, inherent in this process is the requirement for the omnipresent assistance of the administration.

Generally there are a large number of policies that a course director should be familiar with, including those that address student cognitive evaluation and student non-cognitive evaluation. Additional policies that should be included are a leave of absence policy, dismissal policies, academic jeopardy policies, copyright and fair use policies, examination and quiz administration policies, parking regulations, grievance procedures, etc. In the health professions, the manual should also include campus wide policies that address immunizations, HIV policy, exposure to infectious agents, as well as policies regarding the need for students to be trained in basic and advanced cardiac life support.

It is the responsibility of the administration to update the course director's manual annually or when a new policy is added or when an old policy is amended or deleted. The administration should stand ready to help course directors when they encounter student problems and should clarify possible ambiguities in the interpretation of policies.

At the UAMS College of Medicine, a medical educator intranet site was established to facilitate dissemination of policy information. This site contains not only the most recent edition of the course director's manual but also information on copyright and fair use policy, tools for more effective teaching, college of medicine medical student competencies (outcome measures). Also included are student expectations and course and clerkship objectives which underpin the student expectations. Additionally, the intranet site contains course/clerkship methods of student remediation for each course/clerkship objective. There are hyperlinks to various medical education sites and to sites that offer free medically related images for the preparation of PowerPoint presentations. The course directors and faculty alike have found the intranet site to be very helpful.

13.10 Copyright, Fair Use and Intellectual Property Rights

It is the responsibility of the administration to inform course directors of the copyright and fair use policies of the school to ensure that attendant laws are not compromised during the preparation and dissemination of teaching materials. It is the responsibility of the course director to inform all course lecturers of the copyright and fair use policy. It is suggested that the course director give each instructor in his/her course a copy of the school's copyright and fair use policy. Copyright and fair use law must not be violated for the penalties for doing so are harsh and severe for the violator and the school. Copyright and fair use issues come up more frequently than you may think. To help the faculty members understand the policies, the UAMS College of Medicine has prepared a summary of copyright and fair use law in the form of a flyer of frequently asked questions which can be accessed on line at:

http://www.library.uams.edu/policy/copyguide.aspx

and
http://www.library.uams.edu/policy/copycom.aspx

Please be aware that copyright and fair use law is subject to change. Accordingly, course directors need to make sure that they are up-to-date on this important issue. The school library is often a good place to start when learning about rules and regulations, as there is usually a staff member who specializes in this topic.

13.11 Lack of Reward

There is no need to belabor this point. In some medical schools, course directors do not receive an acknowledgement or commensurate monetary compensation from the administration that indicates that they are just as vital in the overall mission of the school as research faculty members and clinical faculty. Being a dedicated educator and/or course director is a labor of love. One teaches and directs a course because one is interested in student learning and instilling lifelong learning and critical thinking skills. Be assured that, as an educator, your dedicated efforts to educate health professionals will touch and influence the lives of many more people than will the outcomes of 99% of the basic science research that is so ardently pursued by your colleagues. Feeling rewarded for what you do for your students is rarely immediate. Satisfaction comes when students see you many months or years later and say that they appreciate what you did for them.

14

Special Considerations

There are several topics that do not fit well into the previous areas of discussion that are important considerations for course directors. We will touch briefly on some of those topics here.

14.1 The Integrated Curriculum

In this section we will not discuss the modalities by which students are taught; rather, we will discuss the manner by which material is linked across course boundaries into a coherent curriculum (i.e., curriculum integration).

Medical school curricula differ as to the system that is employed in the delivery of concepts and principles to the students. Some medical schools utilize a discipline-based curriculum wherein each course and clerkship stands alone. In a discipline-based curriculum very little communication occurs between course directors concerning what is being taught in their course or when topics are presented. In a discipline-based curriculum, students must integrate the material taught in multiple courses, and they have to do this on their own for the most part. As one can imagine, there is variability between students in the quality of information integration. This puts students at a disadvantage when taking standardized tests like the USMLE Step-1 examination where they are expected to be proficient in making connections between concepts and principles across multiple disciplines and courses. In a discipline-based curriculum, redundancy and contradiction occur at a frequency that is generally higher than in an integrated curriculum and this often serves as a catalyst for student dissatisfaction.

A curriculum that is integrated in a horizontal (across an academic year) and vertical (across multiple academic years) fashion allows medical students to see the connections between courses, and in the process think critically in a multidisciplinary fashion – just as they must do in a clinical setting.

The integration of medical school curricula lies on a continuum. Harden (2000) has likened the extent to which curricula are integrated to a ladder, with the rungs described from lowest to highest as follows:

- Isolation
- Awareness
- Harmonization
- Nesting
- Temporal co-coordination

B.W. Newton et al., *How to Become an Effective Course Director*,
DOI 10.1007/978-0-387-84905-8_14, © Springer Science+Business Media, LLC 2009

- Sharing
- Correlation
- Complementary
- Multidisciplinary
- Interdisciplinary
- Trans-disciplinary

On the isolation rung, concurrent courses pay no attention to what the other is teaching. This isolation leaves integration to the students. Material is presented in a disparate and temporally asynchronous fashion, which often leads to student frustration. Isolation can be thought of the as the traditional, discipline-based medical school curriculum.

The awareness rung is characterized by the sharing of handouts with the students and the other course directors so the instructors in different courses will be aware of what is being taught in concurrent courses.

During the harmonization phase of the integration process, teachers in the same and different courses communicate about the information they are teaching. This serves as a means of coordination and results in the reduction of unnecessary redundancy in what is being taught.

On the nesting rung of the ladder, individual courses and disciplines become aware of the broader curriculum metrics and course directors relate what they are teaching to these metrics.

On the temporal co-coordination rung of the ladder, each concurrent course is responsible for its own educational program. However, the timing of topics taught in each of the courses is temporally coordinated.

On the sharing rung, two disciplines plan and implement a teaching program in which overlapping concepts and principles are presented to the students.

On the correlation rung of the ladder, each of the disciplines remains more or less discipline-based and an integrated teaching session is introduced as a means of tying the disparate pieces of the curriculum together.

At the complementary program level, integrated sessions become a major component of the curriculum. The focus of teaching may be a clinical case or some other theme.

As we approach the top of the ladder, we encounter the multi-disciplinary rung. Here, a number of subjects are fused into a single course highlighted with a thematic approach with problems and issues being the target of student learning.

The penultimate step is the inter-disciplinary rung. On this rung, there is heightened emphasis on themes as targets of student learning.

The final step of the curriculum integration ladder is the trans-disciplinary rung. The trans-disciplinary focus is not a theme or topic, but rather a defined field of knowledge as reflected in the reality of where and how that field is manifest, i.e., learning about the nature of a field of knowledge as it is currently defined and practiced. It is safe to say that most medical schools have not attained the top rung of the integration ladder.

It is the authors' view that medical school faculty and curriculum committees should strive to move up the integration ladder as a means of promoting contextually efficient learning. The process of climbing the curriculum integration ladder is ultimately dependent on the medical school faculty who have to decide which step of the ascension process they desire to reach.

The degree of course integration a medical school achieves is dependent on a number of complex variables including whether the control of the curriculum is centralized in the dean's office or, conversely, is controlled by the faculty. It is the view of the authors that faculty should ultimately have control of the curriculum through a curriculum committee or other representative body. At the same time, the policies that dictate curriculum management should give the curriculum committee the authority to address some curricular issues directly without the initial consent of the faculty, but with the consent of the dean of the college.

There are various types of integrated medical school curricula. Some schools integrate their basic science curriculum using an organ-system approach. At the UAMS College of Medicine, the freshman and the sophomore curricula are integrated using an organ-system approach. The freshman curriculum begins with a block called "Molecules to Cells", that includes a Medical Biochemistry course and a Cell Biology course. Both courses are taught in a tightly integrated manner using a clinical case for each week of the block and large group problem-based learning (PBL) sessions that include the entire freshman class rather than smaller groups. After the completion of the Cells to Molecules block, the organ systems are presented. During the organ systems portion of the curriculum, Gross Anatomy, Histology and Physiology run concurrently. Neuroscience begins after Histology and Physiology end, and is integrated with the head and neck portion of Gross Anatomy. Examinations occur at the end of each organ system block, or more frequently for large blocks. All examinations are integrated and computer-based, with the exception of Gross Anatomy practical laboratory examinations. Each course confers a letter grade when the course is completed, and if available, each course uses the respective NBME Subject Examination as their final examination.

The first block of the sophomore curriculum is the "Fundamentals Block". During this block the fundamental concepts and principles of each of the sophomore courses are presented. After the Fundamentals Block, the organ-systems are presented. Each of the organ system blocks is presented in the following sequence: normal development; abnormal development; reactive disease, including infectious disease; and finally neoplasia. Like the freshman year, each of the courses confers a letter grade and, if available, each course uses the respective NBME Subject Examination as their final examination. All of the sophomore in-house examinations are computer-based and make use of integrated, NBME format questions. Using computer-based examinations and asking multidisciplinary questions has proven helpful in preparing our students for the USMLE Step 1 Examination.

It is noteworthy that various teaching modalities are used in the basic science years at the UAMS College of Medicine including lecture, laboratory exercises, conferences, small groups, team-based learning (Michaelsen, Knight, Fink 2004), problem-based learning (Barrows 1994; Stinson, Milter 1996), formative and summative objective-structured clinical examinations (OSCE's), computer self-directed learning modules, and simulations using whole and partial body simulators and task trainers. Clearly, the more approaches one utilizes to teach medical students the better, since medical students (indeed all students) have different learning styles

Some medical schools use an integrated curriculum and classic problem-based learning (PBL) as their modality of instruction. The very nature of classic PBL is highly integrated. Students in small groups are given a clinical case to solve and in the process address the underlying basic science concepts and principles that underpin the case. A faculty member is present during the meetings of the small groups and serves to keep the students on the right track without serving as a constant student reference source. The positive spin-off of this process is that medical students learn the important basic science concepts and principles that form the framework for understanding the clinical presentation of the case that they are addressing, i.e., they learn the basic sciences in a clinical context. Learning the basic sciences in this manner affords the students an appreciation of the importance of basic science knowledge as it relates to the signs and symptoms of disease, as well as to therapeutic interventions. The major problem with using the classic PBL format is the large number of faculty required. A large amount of faculty time is needed to train for PBL sessions as well as to facilitate the small group sessions. This fact has been a deterrent for some medical schools in using classical PBL sessions. At the same time, one must consider the issue of senior faculty adapting to a modality of teaching that is learner-centered rather than instructor-centered wherein the instructor is the font of knowledge.

Some medical schools use a hybrid form of curriculum consisting of lecture and PBL or lecture and team-based learning (TBL). In either case, the degree of integration is dependent on the temporal and topical linkage of the material that is taught.

Other medical schools use case-based curricula (Hark, Morrison 200; Hudson, Buckley 2004; Shanley 2007), wherein a clinical case serves as the infrastructure upon which basic science concepts and principles are addressed. In many ways a case-based curriculum is much like a PBL approach except that small groups are generally not used. Once again, the degree of integration is dependent on the extent with which topics, including concepts and principles, are temporally interwoven, with the ultimate goal of forming a coherent curricular fabric.

14.2 Dealing with Courses/Blocks/Modules in an Integrated Curriculum

In addressing the issue of dealing with courses/blocks/modules, the key word is communication; communication between course directors, and communication between course directors and academic administrators. Course directors must realize that their course does not stand alone and is only one thread in the complex weft and weave of curricular fabric, a weave that is ever changing in response to both internal and external forces.

In an integrated curriculum, the academic dean generally oversees the overall running of the curriculum on a day-to-day basis, often with the assistance of a co-ordinator and a faculty member from the Office of Educational Development, or a similar unit. Assignment of responsibility is also dependent on whether each course in the integrated curriculum has maintained its status as a course or whether the courses have been subsumed into a curriculum in which block grades are given rather than individual course grades. The latter is a curriculum that is generally more integrated than one in which each course maintains its identity by conferring individual course grades.

In an integrated curriculum, someone has to have overall authority to resolve conflicts between courses. In integrated curricula in which courses have maintained their identity, it is sometimes difficult to address problems of unnecessary topic redundancy since courses frequently identify identical topics as being in their teaching sphere. It is the responsibility of the course directors to resolve these issues. But if the course directors cannot resolve their differences of opinion, then the issue must be resolved by the academic administrator. The same rule applies when preparing integrated examination questions – once again compromises should be reached. If that is not possible, the academic administrator should make the decision.

Issues between courses not only occur in the realm of topic redundancy but also as to when didactic information should be taught, i.e. scheduling disagreements. Scheduling issues must also be addressed in an atmosphere of cooperation, and if necessary conflicts resolved by the academic administrator.

The day-to-day teaching of blocks in an integrated curriculum requires constant oversight. The academic dean responsible for the curriculum can carry out the oversight, but ideally it should be the responsibility of individual teaching block directors (affectionately referred to as "block-heads"). The selection of the teaching block directors should be from among the course directors, and should be based on the amount of topical material taught in a given block, i.e., the course director who has the most material in a given block should be the block director. The block director should work with all of the other course directors in achieving optimal integration of the material being taught in his/her block. This requires that each of the course directors submit to the block director a comprehensive outline of the material that they are responsible for addressing in a given block. The block director, working with the curriculum coordinator, should look for topic redundancy and temporal problems in the manner in which the block material is being presented to the students. In some cases it may be necessary to review the PowerPoint presentations or those parts of the syllabus that are to be presented in a given teaching block. This may not be possible at times because instructors may not have their visual aids prepared for a given session. Nevertheless, the block director should do his/her best to see that the material for which he/she is responsible is presented in a well-integrated fashion, both topically and temporally.

Invaluable to making sure that basic science information is clinically relevant, is to have one or more physicians act as an advisor to the content being presented. Having an educational-minded physician present provides the basic scientists the validation that they need to make judgments on whether certain material is clinically relevant.

The content and timing of teaching modules should be overseen by the block director to insure that the modules are integrated into the curriculum. In the context of this writing, a teaching module is a teaching modality that generally serves to concretize information taught within a given teaching block. Some teaching modules are computer-based and afford students an opportunity to apply the information they have learned in a clinical case context. Such teaching modules can serve both a formative and a summative function. Whether formative or summative, teaching modules should be well-integrated within the curriculum. Importantly, they should require medical students to solve problems using critical thinking skills as a means of honing their lifelong learning skills.

Finally, the quality of courses, teaching blocks and learning modules should be addressed by student evaluation and by faculty evaluation. Outcome measures should be pre-established, such as NBME Shelf Examination and USMLE Step 1 section scores, to see if the students are learning what is deemed important on national assessments. The block directors, the course directors, the academic dean and the curriculum committee should address areas of deficiency.

14.3 Critical-Thinking, Problem Solving and Lifelong Learning Skills

Teaching medical students (or any student) to think critically as a means of problem solving, and using those skills to become lifelong learners are not just educational catch phrases – they are the overarching goals of education in general. The aforementioned skills are now LCME educational standards used in the accreditation of all U.S. medical schools (LCME, 2008). Accordingly, course directors should utilize teaching modalities in their course that foster such skills. Several manuscripts have been published that offer suggestions on how to get students to think critically and become lifelong learners (Maudsley, Strivens 2000; Mourtos 2003; Whittle, Murdoch-Eaton 2004; Lim, Hsiung, Hales, 2006). The health of the public is ultimately dependent on the success of medical schools in inculcating these essential skills into the skill sets of medical students – our future physicians.

References

Abrahamson, S. (1996) The State of American Medical Education. In: *Essays on Medical Education*. University Press of America, Inc., New York, pp 103-113.

American Board of Internal Medicine. (2001) Project Professionalism. ABIM Foundation. http://www.abim.org/pdf/publications/professionalism.pdf. Accessed 8 June 2008.

Arnold P. Gold Foundation (2008) http://humanism-in-medicine.org/. Accessed 8 June 2008.

Barrows, H.S. (1994) *Practice-Based Learning: Problem-Based Learning Applied to Medical Education*. Southern Illinois University School of Medicine, Illinois.

Branch, W.T., Jr. (2000) Supporting the moral development of medical students. J. Gen. Intern. Med. 15, 503-508.

Burns, E.R. (2006) Learning syndromes afflicting beginning medical students: identification and treatment — reflections after forty years of teaching. Med. Teach. 28, 230-233.

Draves, W.A. (2002) *Teaching Online*. LERN Books, Wisconsin.

Griffith, C.H., III and Wilson, J.F. (2001) The loss of student idealism in the 3rd-year clinical clerkships. Eval. Health Prof. 24, 61-71.

Hafferty, F.W. (2002) What medical students know about professionalism. Mt. Sinai J. Med. 69, 385-397.

Hafferty, F.W. and Franks, R. (1994) The hidden curriculum, ethics teaching, and the structure of medical education. Acad. Med. 69, 861-871.

Harden, R.M. (2000) The integration ladder: a tool for curriculum planning and evaluation. Med. Educ. 34, 551-557.

Hark, L.A. and Morrison, G. (2000) Development of a case-based integrated nutritional curriculum for medical students. Am. J. Clin. Nutr. 72, 890S-897S.

Hiatt, K.M., Menna J.H., Petty, M., Hackler, C., Hester, M., Mui, D., et al. (2006) A novel medical student examination question appeals process: the committee has spoken! Annual Meeting of the International Association of Medical Science Educators, Puerto Rico, July 15-18, 2006. http://www.iamse.org/conf/conf10/assessment_2006.pdf. Accessed 8 June 2008.

Hudson, J.N. and Buckley, P. (2004) An evaluation of case-based teaching: evidence for continuing benefit and realization of aims. Advan. Physiol. Educ. 28, 15-22.

Inui, T.S. (2003) A flag in the wind: educating for professionalism in medicine. AAMC. www.regenstrief.org/bio/professionalism.pdf/download. Accessed 8 June 2008.

Kay, J. (1990) Traumatic deidealization and the future of medicine. JAMA 263, 572-573.

LCME (2008) Accreditation Standards. http://www.lcme.org/standard.htm. Accessed 8 June 2008.

Lim, R.F., Hsiung, B.C. and Hales, D.J. (2006) Lifelong learning: skills and online resources. Acad. Psychiatry 30, 540-547.

Maudsley, G. and Strivens, J. (2000) Promoting professional knowledge, experiential learning and critical thinking for medical students. Med. Educ. 34, 535-544.

Menna, J.H. and Tank P. (2002) Pre-conference Faculty Development Course: Becoming an Effective Course Director. Sixth Annual Meeting of the International Association of Medical Science Educators, Guadalajara, Mexico, July 20.

Menna, J.H. and Tank P. (2003) Pre-conference Faculty Development Course: Becoming an Effective Course Director. Seventh Annual Meeting of the International Association of Medical Science Educators, Washington, DC., July 19.

Michaelsen, L.K., Knight, A, and Fink, L.D. (2004) *Team-Based Learning: A Transformative Use of Small Groups in College Teaching*. Stylus Publishing, LLC, Sterling, Virginia.

Mourtos, N.J. (2003) Defining, teaching and assessing lifelong learning skills. 33rd ASEE/IEEE Frontiers in Education Conference, November 5-8, Boulder, CO; Session T3B, pp. T3B-14-T3B-19.

Newton, B.W. and Klein, R.M. (2004) Pre-conference Faculty Development Course: Becoming an Effective Course Director. Eighth Annual Meeting of the International Association of Medical Science Educators New Orleans, LA, July 1.

Newton, B.W. and Klein, R.M. (2005) Pre-conference Faculty Development Course: Becoming an Effective Course Director. Ninth Annual Meeting of the International Association of Medical Science Educators, Los Angeles, CA, July 16.

Newton, B.W., Klein, R.M. and Mylona, E. (2006) Pre-conference Faculty Development Course: Becoming an Effective Course Director. Tenth Annual Meeting of the International Association of Medical Science Educators, San Juan, Puerto Rico, July 16.

Newton, B.W., Barber, L., Clardy, J. Cleveland, E. and O'Sullivan, P. (2008) Is there hardening of the heart during medical school? Acad. Med. 83, 244-249.

Papadakis, M.A., Hodgson, C.S., Teherani, A. and Kohatsu, N.D. (2004) Unprofessional behavior in medical school is associated with subsequent disciplinary action by a state medical board. Acad. Med. 79, 244-249.

Papadakis, M.A., Teherani, A., Banach, M.A., Knettler, T.R., Rattner, S.L., et al. (2005) Disciplinary action by medical boards and prior behavior in medical school. NEJM 353, 2673-2682.

Pellegrino, E.D. (2002) Professionalism, profession and virtues of the good physician. Mt. Sinai J. Med. 69, 378-384.

Shanley, P.F. (2007) Viewpoint - leaving the "empty glass" of problem-based learning behind: new assumptions and a revised model for case study in preclinical medical education. Acad. Med. 82, 479-485.

Stinson, J.E. and Milter, R.G. (1996) Problem-based learning in business education: curriculum design and implementation issues. New Directions Teach. Learn. 68, 33-42.

Whittle, S.R. and Murdoch-Eaton, D.G. (2004) Lifelong learning skills: how experienced are students when they enter medical school? Med. Teach. 26, 576-578.

Printed in the United States